Laboratory Manual

Essentials of Biology

Fourth Edition

Sylvia S. Mader

McGraw Hill Education

LABORATORY MANUAL TO ACCOMPANY ESSENTIALS OF BIOLOGY, FOURTH EDITION

Published by McGraw-Hill Education, 2 Penn Plaza, New York, NY 10121. Copyright © 2015 by McGraw-Hill Education. All rights reserved. Printed in the United States of America. Previous editions © 2012, 2010, and 2007. No part of this publication may be reproduced or distributed in any form or by any means, or stored in a database or retrieval system, without the prior written consent of McGraw-Hill Education, including, but not limited to, in any network or other electronic storage or transmission, or broadcast for distance learning.

Some ancillaries, including electronic and print components, may not be available to customers outside the United States.

This book is printed on acid-free paper.

Printed in the United States of America

3 4 5 6 7 8 9 10 QVS/QVS 20 19 18 17 16 15

ISBN 978-0-07-768181-4
MHID 0-07-768181-9

Senior Vice President, Products & Markets: *Kurt L. Strand*
Vice President, General Manager, Products & Markets: *Marty Lange*
Vice President, Content Production & Technology Services: *Kimberly Meriwether David*
Managing Director: *Michael S. Hackett*
Director, Biology: *Lynn Breithaupt*
Brand Manager: *Eric Weber*
Director of Development: *Rose M. Koos*
Senior Marketing Manager: *Chris Loewenberg*
Director, Content Production: *Terri Schiesl*
Content Project Manager: *Sherry Kane*
Senior Buyer: *Sandy Ludovissy*
Senior Designer: *Laurie B. Janssen*
Cover Image: *© FLPA/Bob Gibbons/age fotostock*
Senior Content Licensing Specialist: *Lori Hancock*
Art Studio and Compositor: *Electronic Publishing Services Inc., NYC*
Typeface: *11/13 Utopia Std*
Printer: *Quad Graphics*

Some of the laboratory experiments included in this text may be hazardous if materials are handled improperly or if procedures are conducted incorrectly. Safety precautions are necessary when you are working with chemicals, glass test tubes, hot water baths, sharp instruments, and the like, or for any procedures that generally require caution. Your school may have set regulations regarding safety procedures that your instructor will explain to you. Should you have any problems with materials or procedures, please ask your instructor for help.

The Internet addresses listed in the text were accurate at the time of publication. The inclusion of a website does not indicate an endorsement by the authors or McGraw-Hill Education, and McGraw-Hill Education does not guarantee the accuracy of the information presented at these sites.

Contents

Laboratory	Title	Page	Text Chapter Reference
Unit I The Cell			
1	Scientific Method	1	1
2	Measuring with Metric	9	1
3	Microscopy	19	4
4	Cell Structure and Function	31	4, 5
5	Enzymes	45	5
6	Photosynthesis	53	6
Unit II Genetics			
7	Cellular Reproduction	63	8
8	Sexual Reproduction	73	9
9	Patterns of Inheritance	83	10
10	DNA Biology and Technology	97	11, 12
11	Genetic Counseling	111	13
Unit III Evolution and Diversity of Life			
12	Evidences of Evolution	125	14
13	Microbiology	143	17, 18
14	Plant Evolution	163	18
15	Plant Anatomy and Growth	181	20
16	Animal Evolution	193	19
Unit IV Animal Structure and Function			
17	Basic Mammalian Anatomy I	211	23, 24, 27
18	Chemical Aspects of Digestion	225	24, 25
19	Energy Requirements and Ideal Weight	235	25
20	Basic Mammalian Anatomy II	251	23, 24, 27
21	The Nervous System and Senses	267	27, 28
Unit V Ecology			
22	Effects of Pollution on Ecosystems	287	32

The Virtual Lab Experience!

Based on the same world-class super-adaptive technology as LearnSmart™, McGraw-Hill's LearnSmart Labs™ is a must-see, outcomes-based lab simulation. It assesses a student's knowledge and adaptively corrects deficiencies, allowing the student to learn faster and retain more knowledge with greater success.

First, a student's knowledge is adaptively leveled on core learning outcomes.

Then, a simulated lab experience requires the student to think and act like a scientist. The student is allowed to make mistakes— a powerful part of the learning experience!

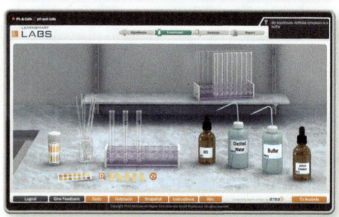

A virtual coach provides subtle hints when needed; asks questions about the student's choices; and allows the student to reflect upon and correct those mistakes.

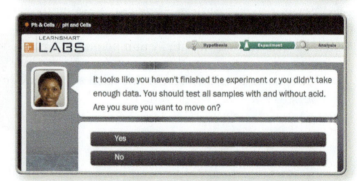

Whether your need is to overcome the logistical challenges of a traditional lab, provide better lab prep, improve student performance, or make your online experience one that rivals the real world, LearnSmart Labs accomplishes it all.

Learn more at www.LearnSmartAdvantage.com

Preface

To the Instructor

The 22 laboratory sessions in this manual have been designed to introduce beginning students to the major concepts of biology, while keeping in mind minimal preparation for sequential laboratory use. The laboratories are coordinated with *Essentials of Biology*, a general biology text that covers all fields of biology. In addition, this Laboratory Manual can be adapted to a variety of course orientations and designs. There are a sufficient number of laboratories and exercises within each lab to tailor the laboratory experience as desired. Then, too, many exercises may be performed as demonstrations rather than as student activities, thereby shortening the time required to cover a particular concept.

Laboratory Resource Guide

The Laboratory Resource Guide, an essential aid for instructors and laboratory assistants, free to adopters of the *Essentials of Biology*, is online at **www.mhhe.com/maderessentials4.** The answers to the Laboratory Review questions are in the Laboratory Resource Guide.

Customized Editions

With McGraw-Hill Create™ you can easily rearrange the labs in this manual and combine them with content from other sources, including your own syllabus or teaching notes. Go to www.mcgrawhillcreate.com to learn how to create your own version of the *Essentials of Biology Laboratory Manual.*

The Exercises

All exercises have been tested for student interest, preparation time, estimated time of completion, and feasibility. The following features are particularly appreciated by adopters:

Integrated opening: Each laboratory begins with a list of Learning Outcomes organized according to the major sections of the laboratory. The major sections of the laboratory are numbered on the opening page and in the laboratory text material. This organization will help students better understand the goals of each laboratory session.

Self-contained content: Each laboratory contains all the background information necessary to understand the concepts being studied and to answer the questions asked. This feature will reduce student frustration and increase learning.

Scientific process: All laboratories stress the scientific process, and many opportunities are given for students to gain an appreciation of the scientific method. The first laboratory of this edition explicitly explains the steps of the scientific method and gives students an opportunity to use them.

Student activities: Sequentially numbered steps guide students as they perform each activity. Some student exercises are Observations (designated by a tan bar) and some are Experimental Procedures (designated by a blue bar). A time icon appears whenever a procedure requires a period of time before results can be viewed.

Live materials: Although students work with living material during some part of almost all laboratories, the exercises are designed to be completed within one laboratory session. This facilitates the use of the manual in multiple-session courses.

Virtual labs: McGraw-Hill's LearnSmart Labs™ are available where noted to accompany many labs in this manual. These outcomes-based lab simulations assess a student's knowledge and adaptively correct deficiencies, allowing the student to learn faster and retain more knowledge with greater success.

Laboratory safety: Laboratory safety, a listing on the website www.mhhe.com, will assist instructors in making the laboratory experience a safe one. Throughout the laboratories, safety precautions specific to an activity are highlighted and identified by a caution symbol.

Improvements This Edition

In general, laboratories have been revised to (1) improve the Introduction so that it becomes an integral part of the laboratory experience; (2) improve the Laboratory Review so that questions better reflect the Learning Outcomes; and (3) use inexpensive plastic transfer pipets when doing exercises that require the use of chemical solutions and test tubes.

Laboratory 1 Scientific Method The laboratory was rewritten to improve the flow and better emphasize the scientific method.

Laboratory 2 Measuring with Metric At adopters' request, study of, and practice with, the metric system has been separated from microscopy and now is presented in a separate laboratory.

Laboratory 7 Cellular Reproduction By following the suggestions of adopters, a laboratory has been created that gives students an exercise to do as they learn the phases of mitosis.

Laboratory 8 Sexual Reproduction In separating meiosis from mitosis, students will engage in an exercise that allows them to see how meiosis introduces genetic variations in the gametes.

Laboratory 9 Patterns of Inheritance The laboratory was rewritten to streamline the experiments to provide students with the opportunity to do hands-on genetics experiments.

Laboratory 12 Evidences of Evolution This laboratory now offers an engaging exposure to plant, animal, and primate fossils. Molecular data to support evolutionary principle were revised.

Laboratory 14 Plant Evolution This laboratory is now based on the latest evolutionary tree of plants and begins with a look at charophytes.

Laboratory 16 Animal Evolution This laboratory includes the newest evolutionary tree of animals and has been shortened to permit use in a single laboratory period.

Laboratory 19 Energy Requirements and Ideal Weight This relevant laboratory is new to the lab manual and provides an easy way for students to decide how best to achieve and maintain an ideal weight.

Laboratory 20 Basic Mammalian Anatomy II This laboratory now begins with coverage of the cardiovascular system.

Laboratory 21 The Nervous System and Senses This laboratory was reorganized to include the spinal cord under the Central Nervous System. The Peripheral Nervous System is better explained and reviewed.

To the Student

Special care has been taken in preparing the *Essentials of Biology Laboratory Manual* to enable you to **enjoy** the laboratory experience as you **learn** from it. The instructions and discussions are written clearly so that you can understand the material while working through it. Student aids are designed to help you focus on important aspects of each exercise. Student learning aids are carefully integrated throughout this manual:

The Learning Outcomes set the goals of each laboratory session and help you review the material for a laboratory practical or any other kind of exam. The major sections of each laboratory are numbered, and the Learning Outcomes are grouped according to these topics. This system allows you to study the chapter in terms of the outcomes presented.

The Introduction to the laboratory reviews necessary background information required for comprehending the work you will be doing during the laboratory session. Also, specific information provided before each of the Observations and Experimental Procedures will assist you in being engaged while you do these activities.

The Observations and Experimental Procedures require your active participation. Space is provided in tables for you to record the results and conclusions of observations and experiments. Space is also provided for you to answer questions pertaining to these activities.

      Observation: An activity in which models, slides, and preserved or live organisms are observed to achieve a learning outcome.

      Experimental Procedure: An activity in which a series of steps uses laboratory equipment to gather data and come to a conclusion.

If you are asked to formulate explanations at the end of an Observation or Experimental Procedure, you should be sure you are truly writing an explanation or conclusion and not just restating the observations made. To do so, you will need to synthesize information from a variety of sources including:

1. Your experimental results and/or the results of other groups in the class. If your data are different from those of other groups in your class, do not erase your answer; add the other groups' answers in parentheses.
2. Your knowledge of underlying principles. Obtain this information from the laboratory Introduction or the appropriate section of the laboratory and from the corresponding chapter of your text.
3. Your understanding of how the experiment was conducted and/or the materials used. (If by chance the results seem inappropriate, it's possible the ingredients were contaminated or you misunderstood the directions. If this occurs, consult with other students and your instructor to see if you should repeat the experiment.)

The Laboratory Review   Each laboratory ends with approximately 10 to 12 short-answer questions that will help you determine if you have accomplished the learning outcomes for the laboratory. The answers to these questions are found in the *Laboratory Resource Guide.*

Color Bars, Time Icon, and Safety Boxes

Observations are identified by the color bar shown to the left. Whenever you see this color, you know that the activity will require you to make careful observations and answer questions about these activities.

Experimental Procedures are identified by the color bar shown to the left. Whenever you see this color, you know that the activity will require you to use laboratory equipment to perform an experiment and answer questions about this experiment.

 A time icon is used to designate when time is needed for a reaction to occur. You may be asked to start these activities at the beginning of the laboratory, proceed to other activities, and return to these when the designated time is up.

⚠️ **A safety icon** throughout the manual alerts you to any specific activity that requires a cautionary approach. Read these boxes, and follow the advice given in the box and/or your instructor when performing the activity.

The Laboratory Review is a set of questions covering the day's work. Do all the review questions as an aid to understanding the laboratory. Your instructor may require you to hand in these questions for credit.

Laboratory Preparation

It will be very helpful to you to read the entire laboratory chapter before coming to lab. **Study** the introductory material and the Observations and Experimental Procedures so you know ahead of time what you will be doing that week and how it correlates with the lecture material. If necessary, to obtain a better understanding, read the corresponding chapter in your text. If your text is *Essentials of Biology* by Sylvia S. Mader, the "text chapter reference" column in the table of contents at the beginning of the *Essentials of Biology Laboratory Manual* lists the corresponding chapter in the text.

Student Feedback

If you have any suggestions for how this Laboratory Manual could be improved, you can send your comments to:

The McGraw-Hill Companies
Product Development—General Biology
501 Bell St.
Dubuque, Iowa 52001

Acknowledgments

We gratefully acknowledge the following reviewers for their assistance in the development of this Laboratory Manual.

Cynthia Anderson
Georgia Military College

Bonnie Arons-Polan
Endicott College

Jack Brook
Mt. Hood Community College

Heidi Bulfer
Colby Community College

Cesar A. Castillo
Queens College, CUNY

Consuella A. Davis
Holmes Community College

Vivian Elder
Ozarks Technical Community College

Melissa Greene
Northwest Mississippi Community College

Maria Gomez
Nicholls State University

Quentin Hays
Eastern New Mexico University

Richard Kerby
Georgia Military College

Martin Matute
University of Arkansas

Esther Muehlbauer
Queens College, CUNY

Kimberly Noice
Richmond Community College

Mohammed Nur-E-Kamal
Medgar Evers College, CUNY

Chris Perry
College of the Albemarle

Carol Phillips
Pamlico Community College

Mary Leigh Poole
Holmes Community College

Adele Register
Rogers State University

Lisa Strong
Northwest Mississippi Community College

Shervia Taylor
Southern University

Anthony Udeogalanya
Medgar Evers College, CUNY

1

Scientific Method

Learning Outcomes

Introduction
- In general, describe pillbug external anatomy and lifestyle.

1.1 Using the Scientific Method
- Outline the steps of the scientific method.
- Distinguish among observations, hypotheses, conclusions, and theories.

1.2 Observing a Pillbug
- Observe and describe the external anatomy of a pillbug, *Armadillidium vulgare.**
- Observe and describe how a pillbug moves.

1.3 Formulating Hypotheses
- Formulate a hypothesis based on appropriate observations.

1.4 Performing the Experiment and Coming to a Conclusion
- Design an experiment that can be repeated by others.
- Reach a conclusion based on observation and experimentation.

Introduction

This laboratory will provide you with an opportunity to use the scientific method in the same manner as scientists. Today your subject will be the pillbug, *Armadillidium vulgare*, a type of crustacean that lives on land.

Pillbugs have an exoskeleton consisting of overlapping "armored" plates that make them look like little armadillos. As pillbugs grow, they molt (shed the exoskeleton) four or five times during a lifetime. A pillbug can roll up into such a tight ball that its legs and head are no longer visible, earning it the nickname "roly-poly." They have three body parts: head, thorax, and abdomen. The head bears compound eyes and two pairs of antennae. The thorax bears pairs of walking legs; gills are located at the top of the first five pairs. The gills must be kept slightly moist, which explains why pillbugs are usually found in damp places. The final pair of appendages, the uropods, which are sensory and defensive in function, project from the abdomen of the animal.

Pillbugs on leaf

Pillbugs are commonly found in damp leaf litter, under rocks, and in basements or crawl spaces under houses. Following an inactive winter, pillbugs mate in the spring. Several weeks later, the eggs hatch and remain for six weeks in a brood pouch on the underside of the female's body. Once they leave the pouch,

---

*The garden snail, *Helix aspersa,* or the earthworm, *Lumbricus terrestris,* can be substituted as desired.

they eat primarily dead organic matter, including decaying leaves. Therefore, they are easy to find and to maintain in a moist terrarium with leaf litter, rocks, and wood chips. You are encouraged to collect some for your experiment. Since they live in the same locations as snakes, be careful when collecting them.

1.1 Using the Scientific Method

Some scientists work alone, but often scientists belong to a community of scientists who are working together to study some aspect of the natural world. For example, many scientists from different institutions work together to study the AIDS virus (Fig. 1.1).

Figure 1.1 Scientists work together.
Robert Gallo and his colleagues do research on how viruses, such as the AIDS virus, invade humans.

You will share your study of pillbugs with the other members of the class.

Even though the methodology can vary, scientists often use the **scientific method** (Fig. 1.2) when doing research. The scientific method involves the following steps.

Making observations. Observations help scientists begin their study of a particular topic.

To learn about pillbugs you will visually observe one. You could also do a Google search of the Web or talk to someone who has worked with pillbugs for a long time.

Why does the scientific method begin with observations? _____

_____

Formulating a hypothesis. Based on their observations, scientists come to a tentative decision, called a hypothesis, about their topic. Formulating hypotheses helps scientists decide how an experiment will be conducted.

Based on your observations you might hypothesize that a pillbug will be attracted to juices.

Now you know what you will actually do. What is the benefit of formulating a hypothesis? _____

_____

Testing the hypothesis involves deciding on an **experimental variable,** that part of the experiment that changes. The dependent variable changes as the experimental variable changes.

You could decide to expose the pillbug to a variety of juices, such as apple juice, orange juice, and pineapple juice.

A well-designed experiment must have a **negative control**—that is, a sample or event—that is not exposed to the testing procedure. If the negative control and the test sample produce the same results, either the procedure is flawed or the hypothesis is false.

Water can substitute for fruit juice and be the control in your experiment.

Scientists call the results of their experiments the **data.** It is very important for scientists to keep accurate records of all their data.

> You will record your data in a table that can be easily examined by another person.

When another person repeats the same experiment, and the data are the same, both experiments have merit. Why must a scientist keep a complete record of an experiment? _____

Coming to a conclusion. Scientists come to a conclusion as to whether their data support or do not support the hypothesis.

> If a pillbug is attracted to fruit juice, your hypothesis is supported. If the pillbug is not attracted to fruit juice, your hypothesis is not supported.

A scientist never says that a hypothesis has been proven true because, after all, some future knowledge might have a bearing on the experiment. What is the purpose of the conclusion? _____

Developing a scientific theory. A *theory* in science is an encompassing conclusion based on many individual conclusions in the same field. For example, the gene theory states that organisms inherit coded information that controls their anatomy, physiology, and behavior. It takes many years for scientists to develop a theory and, therefore, we will not be developing any theories today. How is a scientific theory different from a conclusion? _____

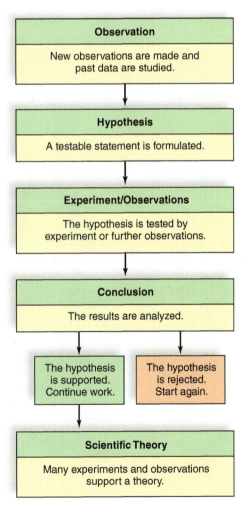

Figure 1.2   Flow diagram for the scientific method.
Often, scientists use this methodology to come to conclusions and develop theories about the natural world. Scientists often choose to retest the same hypothesis, or test a related hypothesis, before arriving at a conclusion.

1.2 Observing a Pillbug

Wash your hands before and after handling pillbugs. Please handle them carefully so they are not crushed. When touched, they roll up into a ball or "pill" shape as a defense mechanism. They will soon recover if left alone.

Observation: Pillbug's External Anatomy

Obtain a pillbug that has been numbered with white correction fluid or tape tags. Put the pillbug in a small glass or plastic dish to keep it contained.

1. Examine the exterior of the pillbug with the unaided eye and with a magnifying lens or dissecting microscope.

   - How can you recognize the head end of a pillbug? _____

   - How many segments and pairs of walking legs are in the thorax? _____

   - The abdomen ends in uropods, appendages with a sensory and defense function. (Females have leaflike growths at the base of some legs where developing eggs and embryos are held in pouches.)

2. In the following space, draw an outline of your pillbug (at least 7 cm long). Label the head, thorax, abdomen, antennae, eyes, uropods, and one of the seven pairs of legs.

3. Draw a pillbug rolled into a ball.

1. Watch a pillbug's underside as the pillbug moves up a transparent surface, such as the side of a graduated cylinder or beaker.

    a. Describe the action of the feet and any other motion you see. _____

    b. Allow a pillbug to crawl on your hand. Describe how it feels. _____

    c. Does a pillbug have the ability to move directly forward? _____

    d. Do you see evidence of mouthparts on the underside of the pillbug? _____

2. As you watch the pillbug, identify

    a. the anatomical parts that allow a pillbug to identify and take in food. _____

    b. behaviors that will help the pillbug acquire food. For example, is the ability of the pillbug to move directly forward a help in acquiring food? How or how not? _____

    What other behaviors allow a pillbug to acquire food? _____

    c. a behavior that helps a pillbug avoid dangerous situations. _____

    If a pillbug rolls up into a ball, wait a few minutes and it may uncurl itself.

3. Measure the speed of three pillbugs.

    a. Place each pillbug on a metric ruler, and use a stopwatch to measure the number of seconds (sec) it takes for the pillbug to move several centimeters (cm). Quickly record here the number of cm moved and the time in sec.

    pillbug 1 _____

    pillbug 2 _____

    pillbug 3 _____

    b. Knowing that 10 millimeters (mm) are in a cm, convert the number of cm traveled to mm and record this in the second column of Table 1.1. Record the total time taken in the third column.

    c. Use the space above to calculate the speed of each pillbug in mm/sec and record the speed for each pillbug in the last column of Table 1.1.

    d. Average the speed for your three pillbugs. (Since you have already calculated the mm/sec for each pillbug, it is only necessary to take an average of the mm moved.) Record the average speed of pillbug motion in Table 1.1.

    When you conduct the experiment in Section 1.4 you will have to be patient with your pillbug as it moves toward or away from a substance.

Table 1.1	Pillbug Speed		
Pillbug	**Millimeters (mm) Traveled**	**Time (sec)**	**Speed (mm/sec)**
1			
2			
3			
		Average speed:	

1.3   Formulating Hypotheses

You will be testing whether pillbugs are attracted to (move toward and eat), repelled by (move away from), or unresponsive to (don't move away from and do not move toward and eat) particular substances, which are potential foods. If a pillbug simply rolls into a ball, nothing can be concluded, and you may wish to choose another pillbug or wait a minute or two to check for further response.

1. Choose
   a. two dry substances, such as flour, cornstarch, coffee creamer, or baking soda. Fine sand will serve as a control for dry substances. Record your "dry" choices as 1, 2, and 3 in the first column of Table 1.2.
   b. two liquids, such as milk, orange juice, ketchup, applesauce, or carbonated beverage. Water will serve as a control for liquid substances. Record your "wet" choices as 4, 5, and 6 in the first column of Table 1.2.
2. In the second column of Table 1.2, hypothesize how you expect the pillbug to respond to each substance. Use a plus (+) sign if you hypothesize that the pillbug will move toward and eat the substance; a minus (−) sign if you hypothesize that the pillbug will be repelled by the substance; and a zero (0) if you expect the pillbug to show neither behavior.
3. In the third column of Table 1.2 offer a reason for your hypothesis based on your knowledge of pillbugs from the introduction and your examination of the animal.

Table 1.2   Hypotheses About Pillbug's Response to Potential Foods

Substance	Hypothesis About Pillbug's Response	Reason for Hypothesis
1		
2		
3 (control)		
4		
5		
6 (control)		

1.4   Performing the Experiment and Coming to a Conclusion

A good experimental design would be to keep your pillbug in a petri dish to test its reaction to the chosen substances. During your experiment, no substance must be put directly on the pillbug, nor can the pillbug be placed directly onto the substances.

Experimental Procedure: Pillbug's Response to Potential Foods

1. Before testing the pillbug's reaction, fill in the first column of Table 1.3. It will look exactly like the first column of Table 1.2.
2. Since pillbugs tend to walk around the edge of a petri dish, you could put the wet or dry substance there; or for the wet substance you could put liquid-soaked cotton in the pillbug's path.
3. Rinse your pillbug between procedures by spritzing it with distilled water from a spray bottle. Then put it on a paper towel to dry.
4. Watch the pillbug's response to each substance, and record it in Table 1.3, using +, −, or 0 as before.

Table 1.3	Pillbug's Response to Potential Foods	
Substance	Pillbug's Response	Hypothesis Supported?
1		
2		
3	(control)	
4		
5		
6	(control)	

5. Do your results support your hypotheses? Answer yes or no in the last column of Table 1.3.

6. Are there any hypotheses that were not supported by the experimental results (data)? How do such

data give you more insight into pillbug behavior? _____

_____

7. Class Results. Compare your results with those of other students who tested the same substance. Calculate the proportional response to each potential food (%+, %−, %0) and record your calculations in Table 1.4. As a group, your class can decide what proportion is needed to designate this response as typical. For example, if the pillbugs as a whole were attracted to a substance 70% or more of the time, you can call that response the "typical response."

Table 1.4	Pillbug's Response to Potential Foods: Class Results			
Substance	Pillbug's Response			Hypothesis Supported?
1	%+	%−	%0	
2	%+	%−	%0	
3 (control)	%+	%−	%0	
4	%+	%−	%0	
5	%+	%−	%0	
6 (control)	%+	%−	%0	

8. On the basis of the class data, do you need to revise your conclusion for any particular pillbug

response? _____ Scientists prefer to come to conclusions on the basis of many trials.

Why is this the best methodology? _____

_____

9. Did the pillbugs respond to the controls as expected (i.e., did not eat them)? _____ If they

did not respond as expected, what can you conclude about your experimental results? _____

_____

1. What are the essential steps of the scientific method? _____

_____

2. What is a hypothesis? _____

_____

3. Is it sufficient to do a single experiment to test a hypothesis? Why or why not? _____

_____

4. What do you call a sample that goes through all the steps of an experiment but does not contain the factor

being tested? _____

5. What part of a pillbug is for protection, and what does a pillbug do to protect itself? _____

_____

6. State the type of data you used to formulate your hypotheses regarding pillbug reactions toward various

substances. _____

7. Why is it important to test one substance at a time when doing an experiment? _____

_____

Indicate whether statements 8–10 are hypotheses, conclusions, or theories.

8. The data show that vaccines protect people from disease. _____

_____

9. All organisms are made of cells. _____

_____

10. The breastbone of a chicken is proportionately larger than that of any other bird. _____

_____

Essentials of Biology Website

Instructors can find lab prep information and answers to all of the laboratory questions in the Laboratory Resource Guide. *Students* can practice their knowledge with quizzes, animations, flashcards, and much more.

www.mhhe.com/maderessentials4

McGraw-Hill Access Science Website

An online encyclopedia of science and technology that provides information, including videos, that can enhance the laboratory experience.

www.accessscience.com

Scientific Method

2

Measuring with Metric

Learning Outcomes

2.1 Length
- Compare and contrast the metric units for length: meter (m), centimeter (cm), millimeter (mm), micrometer (μm), and nanometer (nm).
- Know the abbreviation for each of these units.
- Convert the metric units for length from one type of unit to another.

2.2 Weight
- Compare and contrast the metric units for weight: kilogram (kg), gram (g), and milligram (mg).
- Know the abbreviation for each of these units.
- Convert the metric units for weight from one type of unit to another.

2.3 Volume
- Compare and contrast the metric units for volume: liter (l) and milliliter (ml).
- Know the abbreviation for each of these units.
- Convert the metric units for volume from one type of unit to another.

2.4 Temperature
- Compare and contrast the Fahrenheit (F) and Celsius (C) temperature scales.
- Know the abbreviation for each of these units.
- Convert one type of temperature into another using a provided equation.

2.5 Summary
- Use the size relationship between metric units in order to carry out conversions.

Introduction

The metric system is the standard system of measurement in the sciences, including biology, chemistry, and physics (Fig. 2.1). It has tremendous advantages because all conversions, whether for volume, mass (weight), or length, are in units of ten. This base-ten system is similar to our monetary system, in which 10 cents equals a dime, 10 dimes equals a dollar, and so on. In this laboratory, you will get experience making measurements of length, volume, mass, and temperature.

Figure 2.1
The metric system is the system of measurement used in scientific laboratories.

2.1 Length

Metric units of length measurement include the **meter (m), centimeter (cm), millimeter (mm), micrometer (μm),** and **nanometer (nm)** (Table 2.1).

Table 2.1	Metric Units of Length Measurement			
Unit	Meters	Centimeters	Millimeters	Relative Size
Meter (m)	1 m	100 cm	1,000 mm	Largest
Centimeter (cm)	0.01 (10^{-2}) m	1 cm	10 mm	
Millimeter (mm)	0.001 (10^{-3}) m	0.1 cm	1.0 mm	
Micrometer (μm)	0.000001 (10^{-6}) m	0.0001 (10^{-4}) cm	0.001 (10^{-3}) mm	
Nanometer (nm)	0.000000001 (10^{-9}) m	0.0000001 (10^{-7}) cm	0.000001 (10^{-6}) mm	Smallest

You will want to know what these abbreviations stand for, so write them out here:

m = _____   μm = _____

cm = _____   nm = _____

mm = _____

How many cm are in a meter? _____   How many mm are in a centimeter? _____

How many μm are in a millimeter? _____   How many nm are in a micrometer? _____

_____

Meter, Centimeter, and Millimeter

Observation: A Meterstick

1. Obtain a meterstick. On one side, find the numbers 1 through 39, which denote inches. One meter equals 39.37 inches; therefore, 1 meter is roughly equivalent to 1 yard.

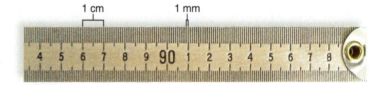

Figure 2.2   The end of a meterstick.

2. Turn the meterstick over and observe the metric subdivisions (Fig. 2.2). How many centimeters are in a meter? _____ The prefix *centi-* means 100. For example, how many cents are in a dollar? _____

3. How many millimeters are in a centimeter? _____ But the prefix *milli-* means 1,000. How many millimeters are in a meter? _____ Obtain a penny and measure its width in terms of mm. _____ Why does it seem preferable to measure a penny in terms of mm? _____

4. Use the meterstick and the method shown in Figure 2.3 to measure the length of two long bones from a disarticulated human skeleton. For example, if the bone measures from the 22 cm mark to the 50 cm mark, the length of the bone is _____ cm. If the bone measures from the 22 cm mark to midway between the 50 cm and 51 cm marks, its length is _____ cm = _____ mm.

5. Record the length of two bones. First bone: _____ cm = _____ mm.

Second bone: _____ cm = _____ mm.

Figure 2.3 Measurement of a long bone.

Lay the meterstick flat on the lab table. Place a long bone next to the meterstick between two pieces of cardboard (each about 10 cm × 30 cm), held upright at right angles to the stick. The narrow end of each piece of cardboard should touch the meterstick. The length between the cards is the length of the bone in centimeters.

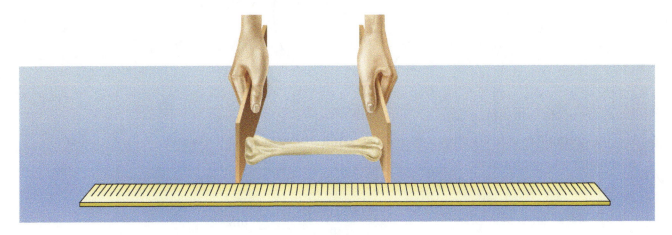

Millimeter, Micrometer, and Nanometer

As you will discover in the next laboratory, the units micrometer (μm) and nanometer (nm) are useful in microscopy because the objects being viewed under the compound light microscope are cells or the contents of cells. Figure 2.4 shows that cells are generally smaller than a millimeter (mm).

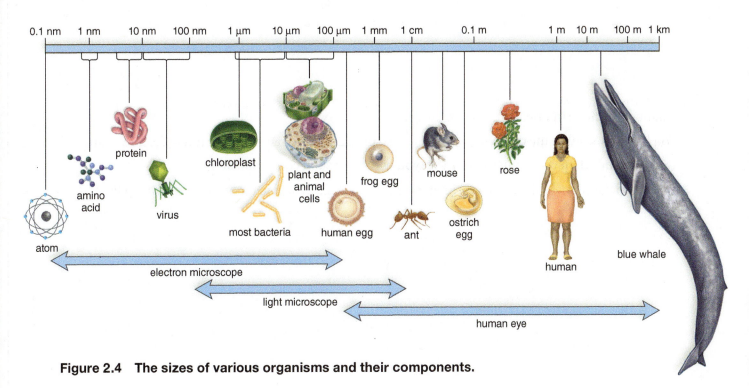

Figure 2.4 The sizes of various organisms and their components.

Laboratory 2 Measuring with Metric **11**

1. Obtain a small metric ruler marked in centimeters (cm) and millimeters (mm). Micrometers (μm) are not on the ruler, and it is necessary to remember that there are 1,000 μm in a mm. Use the ruler to measure the diameter of the circle shown below to the nearest mm.

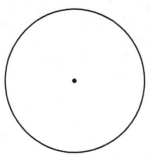

Diameter of circle in mm: _____ mm

2. Use this formula to convert this measurement to μm. Do you expect the answer to be a smaller

number or a larger number? _____

$$\underline{\hspace{1.5cm}} \; mm \times \frac{1,000 \; \mu m}{mm} = \underline{\hspace{1.5cm}} \; \mu m$$

Size of circle in μm: _____ μm

3. Knowing that there are 1,000 nm in a μm, change the formula to convert these μm into nm.

Diameter of circle in nm: _____ nm

4. You have shown that the diameter of this circle is _____ mm = _____ μm = _____ nm.

It is expected when doing microscopy that you will make such conversions.

2.2 Weight

For our purposes we will equate mass with weight. Just as a meter is the basic unit of length, a **kilogram** is the basic unit of mass.

> 1 kilogram (kg) = 1,000 **grams** (g)
> 1 g = 1,000 **milligrams** (mg)

The weight of domestic animals or a human would be given in kilograms, but usually biologists work with the units of grams or milligrams. Using a revision of the formula on page 12 if necessary, do these conversions: 2 g = _____ mg and 0.2 g = _____ mg.

Experimental Procedure: Weight

Examine the balance scales available in your laboratory. They are called balance scales because the object to be weighed is placed in one pan and weights are placed in the other pan. When the pans contain exactly the same mass the beam between them will be in balance. Figure 2.5 shows a typical balance scale you may be using.

Figure 2.5   A balance scale.

1. Use a balance scale to measure the weight of a wooden block small enough to hold in the palm of your hand. Weigh the block to a tenth of a gram. The weight of the wooden block is

   _____ g = _____ mg.

2. Measure the weight of an item small enough to fit inside the opening of a 50 ml graduated cylinder. The item, a(n) _____, is _____ g = _____ mg.

3. A triple beam balance gets its name from its three horizontal beams. If you are going to use a triple beam balance, study it closely (Fig. 2.6). Before making any measurements, clean the weighing pan and move all the weights to the far left. The weights will line up to indicate 0 grams; if they do not, turn the adjustment knob until they do. Measure the weight of an object by placing it in the center of the weighing pan and moving the weights until the beams balance (are straight across). The weight of the object is the sum of the weights on the three beams.

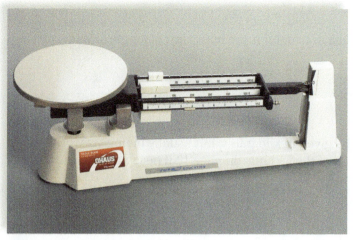

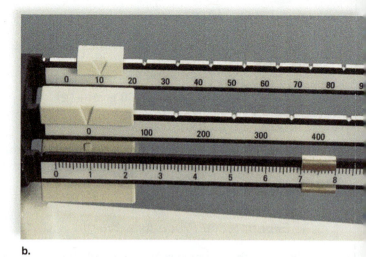

a.

b.

Figure 2.6   A triple beam balance.

a. The entire balance. **b.** Enlargement of triple (three) beams. Each beam has a movable weight. The closest beam weighs to 0.1 g; the middle beam has 100 g graduations; and the farthest beam has 10 g graduations.

If so directed by your instructor, use a triple beam balance to take the weight of one or more of these objects to a tenth of a gram:

Penny   _____

Paper clip   _____

Quarter   _____

2.3   Volume

Two metric units of volume are the **liter (l)** and the **milliliter (ml).** One liter = 1,000 ml.

Experimental Procedure: Volume

1. Volume measurements can be related to those of length. For example, use a millimeter ruler to measure the wooden block used in the previous Experimental Procedure to get its length, width, and depth.

    length = _____ cm; width = _____ cm; depth = _____ cm

    The volume, or space, occupied by the wooden block can be expressed in cubic centimeters (cc or cm³) by multiplying: length × width × depth = _____ cm³. For purposes of this Experimental Procedure, 1 cubic centimeter equals 1 milliliter; therefore, the wooden block you just measured has a volume of _____ ml.

2. In the biology laboratory, liquid volume is usually measured directly in liters or milliliters with appropriate measuring devices. For example, use a 50 ml graduated cylinder to add 20 ml of water to a test tube. First, fill the graduated cylinder to the 20 ml mark. To do this properly, you have to make sure that the lowest margin of the water level, or the **meniscus** (Fig. 2.7), is at the 20 ml mark.

Place your eye directly parallel to the level of the meniscus, and add water until the meniscus is at the 20 ml mark. (Having a dropper bottle filled with water on hand can help you do this.) A large, blank, white index card held behind the cylinder can also help you see the scale more clearly.

3. Pour the 20 ml of water from the graduated cylinder into a test tube. Hypothesize how you could find the total volume of the test

tube. _____

_____

_____

What is the test tube's total volume? _____

_____

4. Fill a 50 ml graduated cylinder with water to about the 20 ml mark. Hypothesize how you could use this setup to calculate the volume of the small object you weighed previously (see step 2, page 13). _____

_____

Now perform the operation you suggested.

The object, _____, has a volume

of _____ ml.

5. Hypothesize how you could determine how many drops from the pipet of the dropper bottle

equal 1 ml. _____

_____

How many drops from the pipet of the dropper

bottle equal 1 ml? _____

6. Some pipets are graduated (Fig. 2.8) and can be filled to a certain level as a way to measure volume directly. Your instructor will demonstrate this. Are pipets customarily used to measure large or small

volumes? _____

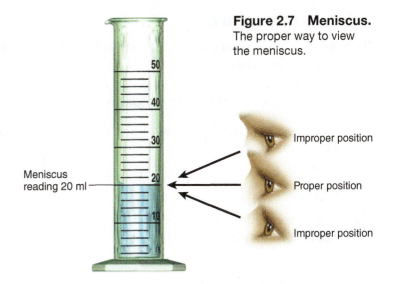

Figure 2.7 Meniscus.
The proper way to view the meniscus.

Meniscus
reading 20 ml

Improper position

Proper position

Improper position

transfer pipet

Figure 2.8 Use of a transfer pipet.
Transfer pipets easily measure and transfer small quantities from stock solutions to test tubes.

2.4　Temperature

There are two temperature scales: the **Fahrenheit (F)** and **Celsius (centigrade, C)** scales (Fig. 2.9). Scientists use the Celsius scale.

Experimental Procedure: Temperature

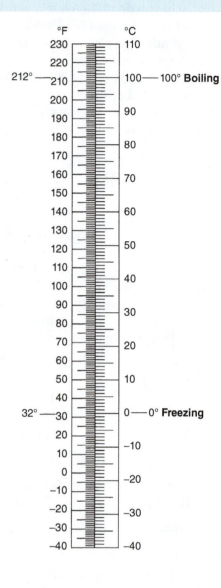

1. Study the two scales in Figure 2.9, and complete the following information:

   a. Water freezes at _____ °F = _____ °C.

   b. Water boils at _____ °F = _____ °C.

2. To convert from the Fahrenheit to the Celsius scale, use the following equation:

   $$°C = (°F - 32°)/1.8$$

   or

   $$°F = (1.8°C) + 32$$

   Human body temperature of 98°F is what temperature on the Celsius scale? _____

3. Record any two of the following temperatures in your lab environment. In each case, allow the Celsius thermometer to remain in or on the sample for 1 minute.

   Room temperature = _____ °C

   Surface of your skin = _____ °C

   Cold tap water in a 50 ml beaker = _____ °C

   Hot tap water in a 50 ml beaker = _____ °C

   Ice water = _____ °C

Figure 2.9　Temperature scales.
The Fahrenheit (°F) scale is on the left, and the Celsius (°C) scale is on the right.

2.5 Summary

Table 2.2 will serve as a summary for this laboratory. In Table 2.2, except for centimeter, the units are all _____ × larger or smaller than the next unit. To know whether to multiply or divide when going from one unit to the next, consider this example:

When you cut an apple into smaller pieces, you end up with more pieces. So, when you convert a gram to milligrams, you multiply the gram by _____. On the other hand, if you cut an apple into 4 pieces, then each piece is .25 of the apple. In the same manner, 1 milligram has to be a portion of a gram. To determine what portion, you divide by _____.

Table 2.2 Units of Metric Measurement	
Length	
Nanometer (nm)	$= 10^{-6}\,m\ (10^{-3}\,\mu m)$
Micrometer (μm)	$= 10^{-6}\,m\ (10^{-3}\,mm)$
Millimeter (mm)	$= 0.001\ (10^{-3})\,m$
Centimeter (cm)	$= 0.01\ (10^{-2})\,m$
Meter (m)	$= 100\ (10^{2})\,cm$
	$= 1{,}000\,mm$
Kilometer (km)	$= 1{,}000\ (10^{3})\,m$
Weight (mass)	
Nanogram (ng)	$= 10^{-9}\,g$
Microgram (μg)	$= 10^{-6}\,g$
Milligram (mg)	$= 10^{-3}\,g$
Gram (g)	$= 1{,}000\,mg$
Kilogram (kg)	$= 1{,}000\ (10^{3})\,g$
Metric ton (t)	$= 1{,}000\,kg$
Volume	
Microliter (μl)	$= 10^{-6}\,l\ (10^{-3}\,ml)$
Milliliter (ml)	$= 10^{-3}\,l$
	$= 1\,cm^{3}\ (cc)$
	$= 1{,}000\,mm^{3}$
Liter (l)	$= 1{,}000\,ml$
Kiloliter (kl)	$= 1{,}000\,l$
Temperature	
Degree	$=$ degree Celsius (°C)

1. What type of measurement is signified by kg? _____ ml? _____ cm? _____
   degrees? _____ μm? _____

2. What type of measurement would utilize a meterstick? _____ a graduated cylinder? _____
   a balance scale? _____

3. If a triple beam balance shows a weight of 100 g plus 10 g plus 1 g, what is the weight of the object? _____
   _____

4. An object is added to a graduated cylinder that holds 250 ml, and the water rises to 300 ml. What is
   the volume of the object in ml? _____ in cm^3? _____

5. Name two units of measurement you expect to use in the next laboratory, which concerns microscopy.
   _____

6. How many micrometers are in a millimeter? _____
   Convert 1.1 mm to μm. _____

7. How many milliliters are in a liter? _____
   Convert 500 ml to liters. _____

8. How many milligrams are in a gram? _____
   Convert 5 g to mg. _____
   _____

9. Convert 1.5 cm to μm. Show your work. _____
   _____

10. A student looking for a shortcut drops an object in a graduated cylinder that contains water to find its
    weight. What's wrong? _____
    _____

Essentials of Biology Website

Instructors can find lab prep information and answers to all of the laboratory questions in the Laboratory Resource Guide. *Students* can practice their knowledge with quizzes, animations, flashcards, and much more.

www.mhhe.com/maderessentials4

McGraw-Hill Access Science Website

An online encyclopedia of science and technology that provides information, including videos, that can enhance the laboratory experience.

www.accessscience.com

LEARNSMART
LABS™

Metric Measurements

3

Microscopy

Introduction

This laboratory examines the features, functions, and use of the compound light microscope and the stereomicroscope. Transmission and scanning electron microscopes are explained, and micrographs produced using these microscopes appear throughout this lab manual. The stereomicroscope and the scanning electron microscope view the surface and/or the three-dimensional structure of an object. The compound light microscope and the transmission electron microscope can view only extremely thin sections of a specimen. If a subject was sectioned lengthwise for viewing, the interior of the projections at the top of the cell, called cilia, would appear in the micrograph (Fig. 3.1). A lengthwise cut through any type of specimen is called a **longitudinal section (l.s.).** On the other hand, if the subject in Figure 3.1 was sectioned crosswise below the area of the cilia, you would see other portions of the interior of the subject. A crosswise cut through any type of specimen is called a **cross section (c.s.).**

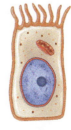

Figure 3.1 Longitudinal and cross sections.
a. Transparent view of a cell. **b.** A longitudinal section shows the cilia at the top of the cell. **c.** A cross section shows only the interior where the cut is made.

a. Cell

b. Longitudinal section (l.s.)

c. Cross section (c.s.)

3.1   Light Microscopes Versus Electron Microscopes

Because biological objects can be very small, we often use a microscope to view them. Many kinds of instruments, ranging from the hand lens to the electron microscope, are effective magnifying devices. A short description of two kinds of light microscopes and two kinds of electron microscopes follows.

Light Microscopes

Light microscopes use light rays passing through lenses to magnify the object. The **stereomicroscope** (binocular dissecting microscope) is designed to study entire objects in three dimensions at low magnification. The **compound light microscope** is used for examining small or thinly sliced sections of objects under higher magnification than that of the stereomicroscope. The term **compound** refers to the use of two sets of lenses: the ocular lenses located near the eyes and the objective lenses located near the object. Illumination is from below, and visible light passes through clear portions but does not pass through opaque portions. To improve contrast, the microscopist uses stains or dyes that bind to cellular structures and absorb light. Photomicrographs, also called light micrographs, are images produced by a compound light microscope (Fig. 3.2*a*).

Figure 3.2   Comparative micrographs of a lymphocyte.
Micrographs of a lymphocyte, a type of white blood cell. **a.** A photomicrograph (light micrograph) shows less detail than a **(b)** transmission electron micrograph (TEM). **c.** A scanning electron micrograph (SEM) shows the cell surface in three dimensions.

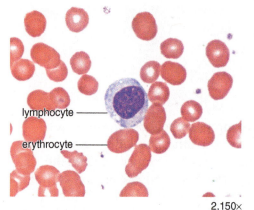

2,150×

a. Photomicrograph, or light micrograph (LM)

2,150×

b. Transmission electron micrograph (TEM)

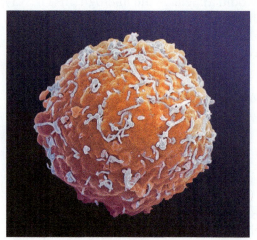

5,000×

c. Scanning electron micrograph (SEM)

Electron Microscopes

Electron microscopes use beams of electrons to magnify the object. The beams are focused on a photographic plate by means of electromagnets. The **transmission electron microscope** is analogous to the compound light microscope. For electron microscopy, the object is ultra-thinly sliced and treated with heavy metal salts to improve contrast. Figure 3.2*b* is a micrograph produced by this type of microscope. The **scanning electron microscope** is analogous to the stereomicroscope. It gives an image of the surface and dimensions of an object, as is apparent from the scanning electron micrograph in Figure 3.2*c*.

The micrographs in Figure 3.2 demonstrate that an object is magnified more with an electron microscope than with a compound light microscope. The difference between these two types of microscopes, however, is not simply a matter of magnification; it is also the electron microscope's ability to show detail. The electron microscope has greater resolving power. **Resolution** is the minimum distance between two objects at which they can still be seen, or resolved, as two separate objects. The use of high-energy electrons rather than light gives electron microscopes a much greater resolving power, since two objects that are much closer together can still be distinguished as separate points. Table 3.1 lists several other differences between the compound light microscope and the transmission electron microscope.

Table 3.1	Comparison of the Compound Light Microscope and the Transmission Electron Microscope	
Compound Light Microscope		**Transmission Electron Microscope**
1. Glass lenses		1. Electromagnetic lenses
2. Illumination by visible light		2. Illumination due to beam of electrons
3. Resolution $\cong$ 200 nm		3. Resolution $\cong$ 0.1 nm
4. Magnifies to 1,500×		4. Magnifies to 100,000×
5. Costs up to tens of thousands of dollars		5. Costs up to hundreds of thousands of dollars

Answer These Questions

- Which two types of microscopes view the surface of an object? _____

  _____

- Which two types of microscopes view objects that have been sliced and treated to improve

  contrast? _____

- Of the microscopes just mentioned, which one resolves the greater amount of detail? _____

  _____

Rules for Microscope Use

Observe the following rules for using a microscope:

1. The lowest-power objective (scanning or low) should be in position, at both the beginning and the end of microscope use.
2. Use only lens paper for cleaning lenses.
3. Do not tilt the microscope because the eyepieces could fall out, or wet mounts could be ruined.
4. Keep the stage clean and dry to prevent rust and corrosion.
5. Do not remove parts of the microscope.
6. Keep the microscope dust-free by covering it after use.
7. Report any malfunctions.

3.2 Stereomicroscope (Binocular Dissecting Microscope)

The **stereomicroscope** (binocular dissecting microscope) allows you to view objects in three dimensions at low magnifications. It is used to study entire small organisms, any object requiring lower magnification, and opaque objects that can be viewed only by reflected light. It is also called a stereomicroscope because it produces a three-dimensional image.

Identifying the Parts

After your instructor has explained how to carry a microscope, obtain a stereomicroscope and a separate illuminator, if necessary, from the storage area. Place it securely on the table (Fig. 3.3). Plug in the power cord, and turn on the illuminator. There is a wide variety of stereomicroscope styles, and your instructor will discuss the specific style(s) available to you. Regardless of style, the following features should be present:

Figure 3.3   Stereomicroscope (binocular dissecting microscope).
Label this microscope with the help of the text material.

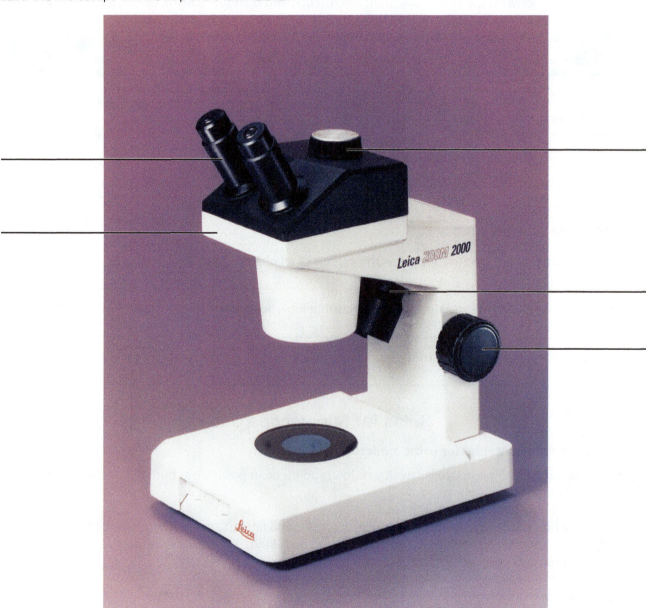

1. **Binocular head:** Holds two eyepiece lenses that move to accommodate for the various distances between different individuals' eyes.
2. **Eyepiece lenses:** The two lenses located on the binocular head. What is the magnification of your eyepieces? _____ Some models have one **independent focusing eyepiece** with a knob to allow independent adjustment of each eye. The nonadjustable eyepiece is called the **fixed eyepiece.**
3. **Focusing knob:** A large, black or gray knob located on the arm; used for changing the focus of both eyepieces together
4. **Magnification changing knob:** A knob, often built into the binocular head, used to change magnification in both eyepieces simultaneously. This may be a **zoom** mechanism or a **rotating lens** mechanism of different powers that clicks into place.
5. **Illuminator:** Used to illuminate an object from above; may be built into the microscope or separate

Locate each of these parts on your stereomicroscope, and label them on Figure 3.3.

Focusing the Stereomicroscope

1. Place a plastomount that contains small organisms in the center of the stage.
2. Adjust the distance between the eyepieces on the binocular head so that they comfortably fit the distance between your eyes. You should be able to see the object with both eyes as one three-dimensional image.
3. Use the focusing knob to bring an organism in the plastomount into focus.
4. Does your microscope have an independent focusing eyepiece? _____ If so, use the focusing knob to bring the image in the fixed eyepiece into focus, while keeping the eye at the independent focusing eyepiece closed. Then adjust the independent focusing eyepiece so that the image is clear, while keeping the other eye closed. Is the image inverted? _____
5. Turn the magnification changing knob, and determine the kind of mechanism on your microscope. A zoom mechanism allows continuous viewing while changing the magnification. A rotating lens mechanism blocks the view of the object as the new lenses are rotated. Be sure to click each lens firmly into place. If you do not, the field will be only partially visible. What kind of mechanism is on your microscope? _____
6. Set the magnification changing knob on the lowest magnification. *Sketch an organism from the plastomount in the following circle as though this represents your entire field of view:*

7. Rotate the magnification changing knob to the highest magnification. *Draw another circle within the one provided to indicate the reduction of the field of view.*
8. Experiment with various objects at various magnifications until you are comfortable with using the stereomicroscope.
9. When you are finished, return your stereomicroscope and illuminator to their correct storage areas.

3.3 Use of the Compound Light Microscope

As mentioned, the name "compound light microscope" indicates that it uses two sets of lenses and light to view an object. The two sets of lenses are the ocular lenses located near the eyes and the objective lenses located near the object. Illumination is from below, and the light passes through clear portions but does not pass through opaque portions of the object. This microscope is used to examine small or thinly sliced sections of objects under higher magnification than would be possible with the stereomicroscope.

Identifying the Parts

Obtain a compound light microscope from the storage area, and place it securely on the table. *Identify the following parts on your microscope, and label them in Figure 3.4.*

Figure 3.4 Compound light microscope.
Compound light microscope with binocular head and mechanical stage. Label this microscope with the help of the text material.

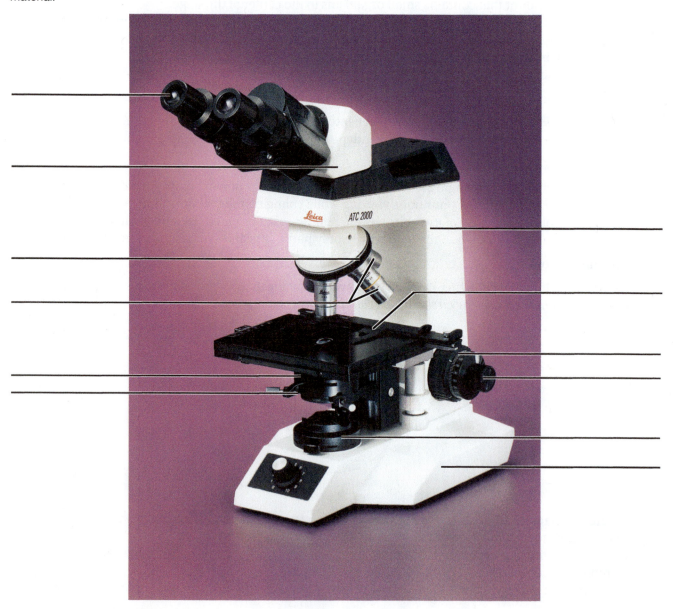

1. **Ocular lenses (eyepieces):** What is the magnifying power of the ocular lenses on your microscope? _____

2. **Body tube:** Holds nosepiece at one end and eyepiece at the other end; conducts light rays

3. **Arm:** Supports upper parts and provides carrying handle. When carrying the microscope, hold the arm with one hand and place the other hand beneath the base.

4. **Nosepiece:** Revolving device that holds objectives

5. **Objective lenses (objectives):**

   a. **Scanning objective:** This is the shortest of the objective lenses and is used to scan the whole slide. The magnification is stamped on the housing of the lens. It is a number followed by an ×. What is the magnifying power of the scanning objective lens on your microscope? _____

   b. **Low-power objective:** This lens is longer than the scanning objective lens and is used to view objects in greater detail. What is the magnifying power of the low-power objective lens on your microscope? _____

   c. **High-power objective:** If your microscope has three objective lenses, this lens will be the longest. It is used to view an object in even greater detail. What is the magnifying power of the high-power objective lens on your microscope? _____

   d. **Oil immersion objective** (on microscopes with four objective lenses): Holds a 95× (to 100×) lens and is used in conjunction with immersion oil to view objects with the greatest magnification.

      Does your microscope have an oil immersion objective? _____ If this lens is available, your instructor will discuss its use when the lens is needed.

6. **Coarse-adjustment knob:** Knob used to bring object into approximate focus; used only with low-power objective

7. **Fine-adjustment knob:** Knob used to bring object into final focus

8. **Condenser:** Lens system below the stage used to focus the beam of light on the object being viewed

9. **Diaphragm** or **diaphragm control lever:** Controls amount of illumination used to view the object

10. **Light source:** An attached lamp that directs a beam of light up through the object

11. **Base:** The flat surface of the microscope that rests on the table

12. **Stage:** Holds and supports microscope slides

13. **Stage clips:** Hold slides in place on the stage

14. **Mechanical stage (optional):** A movable stage that aids in the accurate positioning of the slide.

    Does your microscope have a mechanical stage? _____

15. **Mechanical stage control knobs (optional):** Two knobs usually located below the stage. One knob controls forward/reverse movement, and the other controls right/left movement.

Focusing the Compound Light Microscope—Lowest Power

1. Turn the nosepiece so that the lowest-power objective on your microscope is in straight alignment over the stage.

2. Always begin focusing with the lowest-power objective on your microscope (4× [scanning] or 10× [low power]).

3. With the coarse-adjustment knob, lower the stage (or raise the objectives) until it stops.

4. Place a slide of the letter *e* on the stage, and stabilize it with the clips. (If your microscope has a mechanical stage, pinch the spring of the slide arms on the stage, and insert the slide.) Center the *e* as best you can on the stage, or use the two control knobs located below the stage (if your microscope has a mechanical stage) to center the *e*.

5. Again, be sure that the lowest-power objective is in place. Then, as you look from the side, decrease the distance between the stage and the tip of the objective lens until the lens comes to an automatic stop or is no closer than 3 mm above the slide.

6. While looking into the eyepiece, rotate the diaphragm (or diaphragm control lever) to give the maximum amount of light.

7. Using the coarse-adjustment knob, slowly increase the distance between the stage and the objective lens until the object—in this case, the letter *e*—comes into view, or focus.

8. Once the object is seen, you may need to adjust the amount of light. To increase or decrease the contrast, rotate the diaphragm slightly.

9. Use the fine-adjustment knob to sharpen the focus if necessary.

10. Practice having both eyes open when looking through the eyepiece, as this greatly reduces eyestrain.

Inversion

Inversion refers to the fact that a microscopic image is upside down and reversed.

Observation: Inversion

1. In space 1 provided here, *draw the letter* e *as it appears on the slide (with the unaided eye, not looking through the eyepiece).*

1.	2.

2. In space 2, *draw the letter* e *as it appears when you look through the eyepiece.*

3. What differences do you notice? _____

4. Move the slide to the right. Which way does the image appear to move? _____

5. Move the slide down. Which way does the image move? _____

Focusing the Compound Light Microscope—Higher Powers

Compound light microscopes are **parfocal;** that is, once the object is in focus with the lowest power, it should also be almost in focus with the higher power.

1. Bring the object into focus under the lowest power by following the instructions in the previous section.

2. Make sure that the letter *e* is centered in the field of the lowest objective.

3. Move to the next higher objective (low power [10×] or high power [40×]) by turning the nosepiece until you hear it click into place. Do not change the focus; parfocal microscope objectives will not hit normal slides when changing the focus if the lowest objective is initially in focus. (If you are on low power [10×], proceed to high power [40×] before going on to step 4.)

4. If any adjustment is needed, use only the *fine*-adjustment knob. (*Note:* Always use only the fine-adjustment knob with high power.) On your drawing of the letter *e, draw a circle around the portion of the letter that you are now seeing with high-power magnification.*

5. When you have finished your observations of this slide (or any slide), rotate the nosepiece until the lowest-power objective clicks into place, and then remove the slide.

Total Magnification

Total magnification is calculated by multiplying the magnification of the ocular lens (eyepiece) by the magnification of the objective lens. (The ocular lenses are each 10×.) The magnification of objective lenses is imprinted on the lens casing.

Observation: Total Magnification

In Table 3.2 calculate total magnifications for your microscope according to the ocular lens and the objective lens you are using.

Table 3.2 Total Magnification			
Objective	**Ocular Lens**	**Objective Lens**	**Total Magnification**
Scanning power (if present)			
Low power			
High power			
Oil immersion (if present)			

3.4 Microscopic Observations

When a specimen is prepared for observation, the object should always be viewed as a **wet mount.** A wet mount is prepared by placing a drop of liquid on a slide or, if the material is dry, by placing it directly on the slide and adding a drop of water or stain. The mount is then covered with a coverslip, as illustrated in Figure 3.5. Dry the bottom of your slide before placing it on the stage.

Figure 3.5 Preparation of a wet mount.

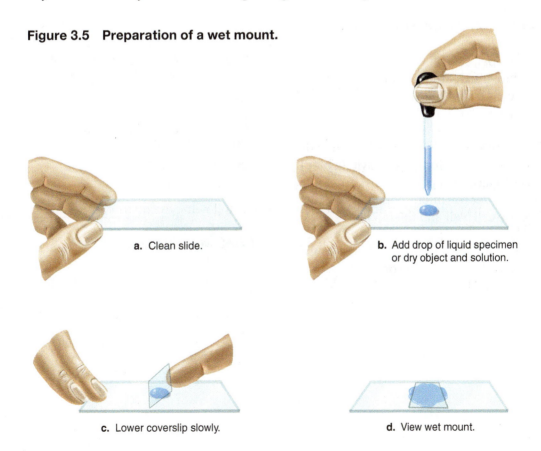

a. Clean slide.

b. Add drop of liquid specimen or dry object and solution.

c. Lower coverslip slowly.

d. View wet mount.

Human Epithelial Cells

Epithelial cells cover the body's surface and line its cavities.

Observation: Human Epithelial Cells

1. Obtain a prepared slide, or make your own as follows:
   a. Obtain a prepackaged flat toothpick (or sanitize one with alcohol or alcohol swabs).
   b. Gently scrape the inside of your cheek with the toothpick, and place the scrapings on a clean, dry slide. Discard used toothpicks in the biohazard waste container provided.
   c. Add a drop of very weak methylene blue or iodine solution, and cover with a coverslip.
2. Observe under the microscope.
3. Locate the nucleus (the central, round body), the cytoplasm, and the plasma membrane (outer cell boundary). *Label Figure 3.6.*
4. Because your epithelial slides are biohazardous, they must be disposed of as indicated by your instructor.

Figure 3.6   Cheek epithelial cells.
Label the nucleus, the cytoplasm, and the plasma membrane.

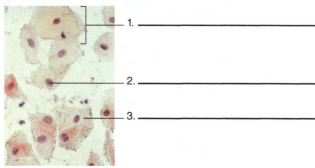

1. _____
2. _____
3. _____

1,000×

Onion Epidermal Cells

Epidermal cells cover the surfaces of plant organs, such as leaves. The bulb of an onion is made up of fleshy leaves.

Observation: Onion Epidermal Cells

1. With a scalpel, strip a small, thin, transparent layer of cells from the inside of a fresh onion leaf.

⚠ **Scalpel** Exercise care when using a scalpel.

2. Place it gently on a clean, dry slide, and add a drop of iodine solution (or methylene blue). Cover with a coverslip.
3. Observe under the microscope.
4. Locate the cell wall and the nucleus. *Label Figure 3.7.*
5. Note some obvious differences between the human cheek cells and the onion cells, and list them in Table 3.3.

Figure 3.7   Onion epidermal cells.
Label the cell wall and the nucleus.

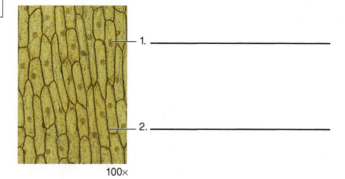

1. _____
2. _____

100×

Table 3.3   Differences Between Human Epithelial and Onion Epidermal Cells

Differences	Human Epithelial Cells (Cheek)	Onion Epidermal Cells
Shape (sketch)		

Euglena

Examination of *Euglena* (a unicellular organism with a flagellum to facilitate movement) will test your ability to observe objects with the microscope and to control illumination to heighten contrast.

Observation: Euglena

1. Make a wet mount of *Euglena* by using a drop of a *Euglena* culture and adding a drop of Protoslo (methyl cellulose solution) onto a slide. The Protoslo slows the organism's swimming.
2. Mix thoroughly with a toothpick, and add a coverslip.
3. Scan the slide for *Euglena:* Start at the upper left-hand corner, and move the slide forward and back as you work across the slide from left to right. The *Euglena* may be at the edge of the slide because they show an aversion to Protoslo.
4. Experiment by using scanning, low-power, and high-power objective lenses; by focusing up and down with the fine-adjustment knob; and by adjusting the light so that it is not too bright.
5. Compare your *Euglena* specimens with Figure 3.8. How do your specimens compare to Figure 3.8? _____

_____

Figure 3.8   *Euglena.*
Euglena is a unicellular, flagellated organism.

200×

Laboratory Review 3

1. Of the three types of microscopes studied, which one best shows the surface of an object? _____
_____

2. Explain the designation "compound light" microscope:

   a. compound _____

   b. light _____

3. What function is performed by the diaphragm of a microscope? _____

4. Briefly describe the necessary steps for observing a slide at low power under the compound light

   microscope. _____

   _____

5. Why is it helpful for a microscope to be parfocal? _____

   _____

6. Why is locating an object more difficult if you start with the high-power objective than with the low-power

   objective? _____

7. How much larger than normal does an object appear with a low-power objective? _____

   _____

8. A virus is 50 nm in size. Would you recommend using a stereomicroscope, a compound light microscope, or

   an electron microscope to see it? _____ Why? _____

9. What types of microscopes might you use to observe the live organisms found in pond water? _____

   _____

10. State two differences between onion epidermal cells and human epithelial cells. _____

    _____

    _____

4

Cell Structure and Function

Learning Outcomes

4.1 Animal Cell and Plant Cell Structure
- Label an animal cell diagram, and state a function for the structures labeled.
- Label a plant cell diagram, and state a function for the structures labeled.
- Use microscopic techniques to observe plant cell structure.

4.2 Diffusion
- Define and describe the process of diffusion as affected by the medium.
- Predict and observe which substances will or will not diffuse across a plasma membrane.

4.3 Osmosis: Diffusion of Water Across Plasma Membrane
- Explain an osmosis experiment based on a knowledge of diffusion principles.
- Define isotonic, hypertonic, and hypotonic solutions, and give examples in terms of NaCl concentrations.
- Predict the effect of different tonicities on animal (e.g., red blood) cells and on plant (e.g., *Elodea*) cells.

4.4 pH and Cells
- Predict the change in pH before and after the addition of an acid to nonbuffered and buffered solutions.
- Suggest a method by which it is possible to test the effectiveness of antacid medications.

Introduction

The molecules we studied in Laboratory 3 are not alive—the basic units of life are cells. The **cell theory** states that all organisms are composed of cells and that cells come only from other cells. While we are accustomed to considering the heart, the liver, or the intestines as enabling the human body to function, it is actually cells that do the work of these organs.

Figure 3.6 shows human cheek epithelial cells as viewed by an ordinary compound light microscope available in general biology laboratories. It shows that the content of a cell, called the **cytoplasm,** is bounded by a **plasma membrane.** The plasma membrane regulates the movement of molecules into and out of the cytoplasm. In this lab, we will study how the passage of water into a cell depends on the difference in concentration of solutes (particles) between the cytoplasm and the surrounding medium or solution. The well-being of cells also depends upon the pH of the solution surrounding them. We will see how a buffer can maintain the pH within a narrow range and how buffers within cells can protect them against damaging pH changes.

Because a photomicrograph shows only a minimal amount of detail, it is necessary to turn to the electron microscope to study the contents of a cell in greater depth. The models of plant and animal cells available in the laboratory today are based on electron micrographs.

> **Planning Ahead** To save time, your instructor may have you start a boiling water bath (page 37) and the potato strip experiment (page 41) at the beginning of the laboratory.

4.1 Animal Cell and Plant Cell Structure

Table 4.1 lists the structures found in animal and plant cells. The **nucleus** in a eukaryotic cell is bounded by a **nuclear envelope** and contains **nucleoplasm. Cytoplasm,** found between the plasma membrane and the nucleus, consists of a gel-like fluid and also the organelles, such as the nucleolus, endoplasmic reticulum, Golgi apparatus, vacuoles and vesicles, lysosomes, peroxisome, mitochondrion, and chloroplast.

Table 4.1	**Eukaryotic Structures in Animal Cells and Plant Cells**	
Name	**Composition**	**Function**
Cell wall*	Contains cellulose fibrils	Provides support and protection
Plasma membrane	Phospholipid bilayer with embedded proteins	Outer cell surface that regulates entrance and exit of molecules
Nucleus	Enclosed by nuclear envelope; contains chromatin (threads of DNA and protein)	Storage of genetic information; synthesis of DNA and RNA
Nucleolus	Concentrated area of chromatin	Produces subunits of ribosomes
Ribosome	Protein and RNA in two subunits	Carries out protein synthesis
Endoplasmic reticulum (ER)	Membranous, flattened channels and tubular canals; rough ER and smooth ER	Synthesis and/or modification of proteins and other substances; transport by vesicle formation
Rough ER	Studded with ribosomes	Protein synthesis
Smooth ER	Lacks ribosomes	Synthesis of lipid molecules
Golgi apparatus	Stack of membranous saccules	Processes, packages, and distributes proteins and lipids
Vesicle/vacuole	Membrane-bounded sac; large central vacuole in plant cells*	Stores and transports substances
Lysosome	Vesicle containing hydrolytic enzymes	Digests macromolecules and cell parts
Peroxisome	Vesicle containing specific enzymes	Breaks down fatty acids and converts resulting hydrogen peroxide to water
Mitochondrion	Bounded by double membrane; inner membrane is cristae	Cellular respiration, producing ATP molecules
Chloroplast*	Membranous grana bounded by double membrane	Photosynthesis, producing sugars
Cytoskeleton	Microtubules, intermediate filaments, actin filaments	Maintains cell shape and assists movement of cell parts
Cilia and flagella	Attachments supported by microtubules	Movement of cell
Centrioles** in centrosome	Microtubule-containing, cylindrically shaped organelle in a structure of complex composition	Centrioles organize microtubules in cilia and flagella; centrosome organizes microtubules in cell

*Plant cells only

**Animal cells only

Study Table 4.1 to determine structures that are unique to plant cells and unique to animal cells, and write them below the examples given.

	Plant Cells	**Animal Cells**
Unique structures:	1. Large central vacuole	1. Small vacuoles
	2. _____	2. _____
	3. _____	

Animal Cell Structure

Label Figure 4.1. With the help of Table 4.1, give a function for these labeled structures.

Structure	Function
Plasma membrane	_____
Nucleus	_____
Nucleolus	_____
Endoplasmic reticulum	_____
Rough ER	_____
Smooth ER	_____
Golgi apparatus	_____
Vesicle	_____
Lysosome	_____
Mitochondrion	_____
Centrioles in centrosome	_____
Cytoskeleton	_____

Figure 4.1 Animal cell structure.

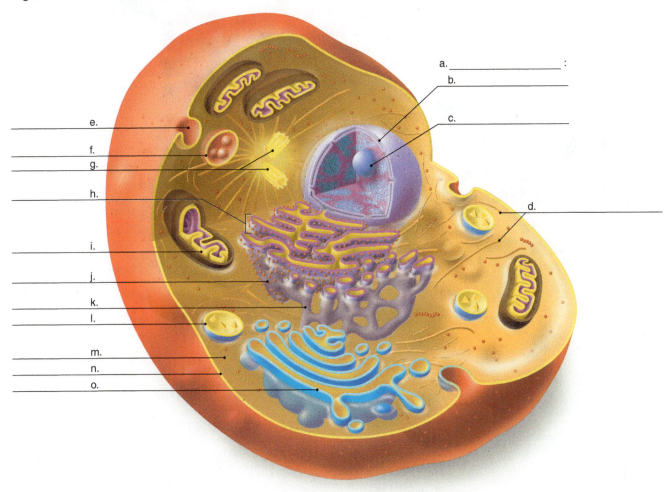

Plant Cell Structure

Label Figure 4.2. With the help of Table 4.1, give a function for each labeled structure unique to plant cells.

Structure   **Function**

Cell wall _____

Central vacuole, large _____

Chloroplast _____

Figure 4.2   Plant cell structure.

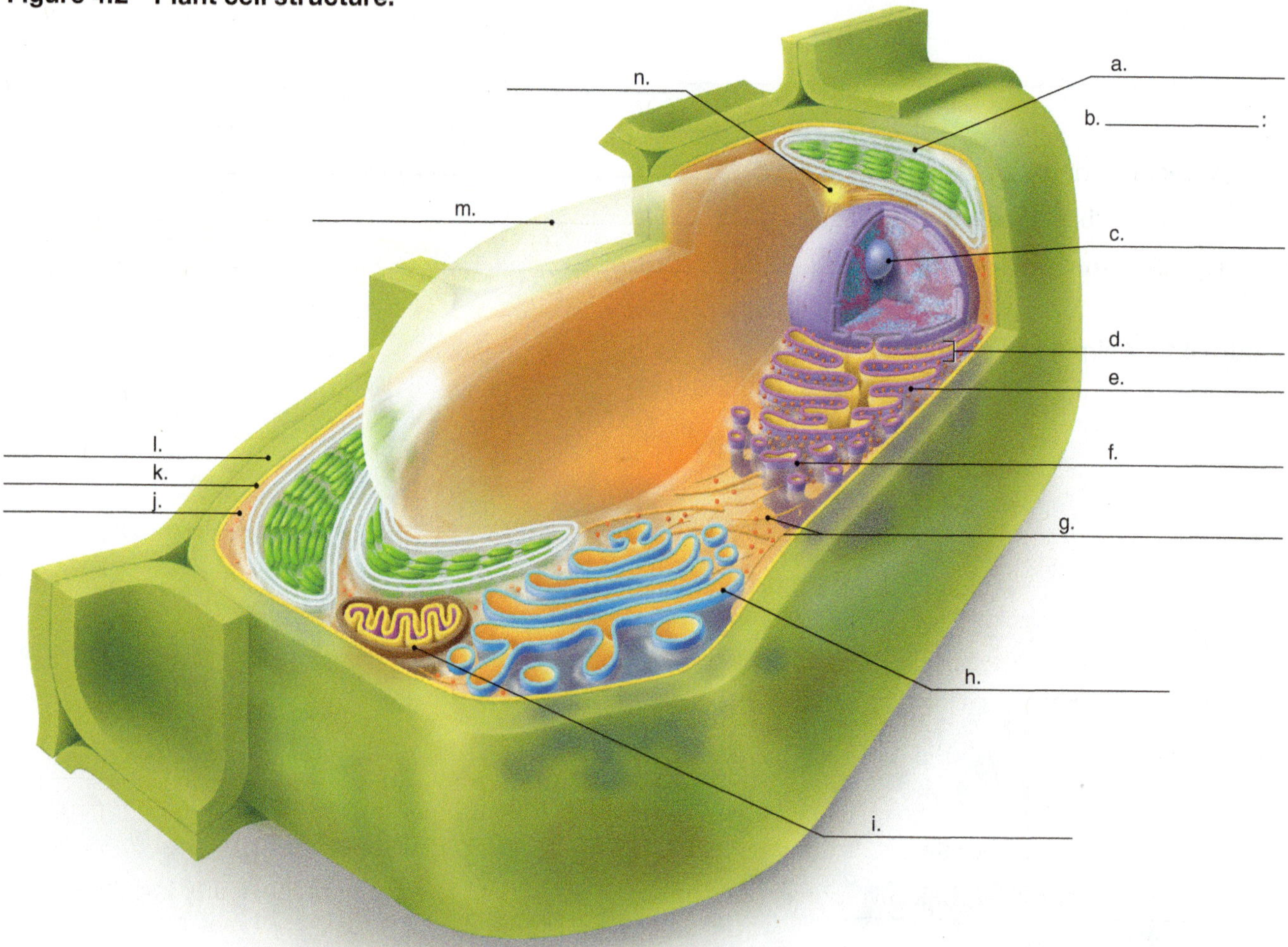

Observation: Plant Cell Structure

1. Prepare a wet mount of a small piece of young *Elodea* leaf in fresh water. *Elodea* is a multicellular, eukaryotic plant found in freshwater ponds and lakes.
2. Have the drop of water ready on your slide so that the leaf does not dry out, even for a few seconds. Take care that the leaf is mounted with its top side up.
3. Using low power bring the leaf surface into focus.
4. Select a cell with numerous chloroplasts for further study, and switch to high power.
5. Carefully focus on the sides of the cell. The chloroplasts appear to be only along the sides of the cell because the large, fluid-filled, membrane-bounded central vacuole pushes the cytoplasm against the cell walls (Fig. 4.3*a*). Then focus on the surface and notice an even distribution of chloroplasts (Fig. 4.3*b*).

Figure 4.3 *Elodea* cell structure.

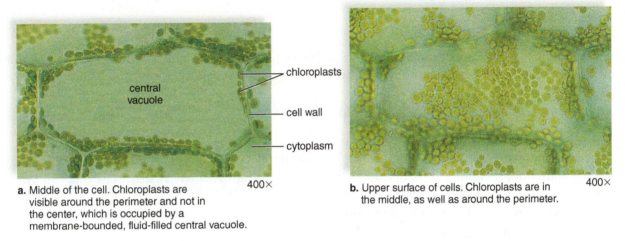

a. Middle of the cell. Chloroplasts are visible around the perimeter and not in the center, which is occupied by a membrane-bounded, fluid-filled central vacuole. 400×

b. Upper surface of cells. Chloroplasts are in the middle, as well as around the perimeter. 400×

6. Can you locate the cell nucleus? _____ It may be hidden by the chloroplasts, but when visible, it appears as a faint, gray lump on one side of the cell.

7. Why can't you see the other organelles featured in Figure 4.2? _____

_____

8. Can you detect movement of chloroplasts in this cell or any other cell? _____ The chloroplasts are not moving under their own power but are being carried by a streaming of the nearly invisible cytoplasm.

9. Save your slide for use later in this laboratory.

4.2 Diffusion

Diffusion is the movement of molecules from a higher to a lower concentration until equilibrium is achieved and the molecules are distributed equally (Fig. 4.4). At equilibrium, molecules may still be moving back and forth, but there is no net movement in any one direction.

Figure 4.4 Process of diffusion.
Diffusion is apparent when dye molecules have equally dispersed.

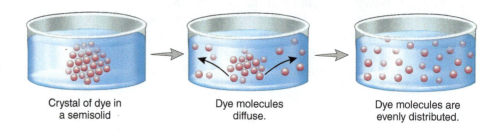

Crystal of dye in a semisolid ·   Dye molecules diffuse.   Dye molecules are evenly distributed.

Diffusion Through a Semisolid

Diffusion of molecules is a general phenomenon in the environment. The speed of diffusion is dependent on such factors as the temperature, the size of the molecule, and the type of medium.

Experimental Procedure: Diffusion in a Semisolid

1. Observe a petri dish containing 1.5% gelatin (or agar) to which potassium permanganate ($KMnO_4$) was added in the center depression.

2. Using the start time posted by your instructor and the time of day now, calculate how long diffusion has been occurring. Length of time has been _____.

> ⚠️ **Potassium permanganate ($KMnO_4$)** $KMnO_4$ is highly poisonous and is a strong oxidizer. Avoid contact with skin and eyes and with combustible materials. If spillage occurs, wash all surfaces thoroughly. $KMnO_4$ will also stain clothing.

3. Using a ruler placed over the petri dish, measure (in mm) the movement of color from the center of the depression outward in one direction: _____ mm.

4. Calculate the speed of diffusion: _____ mm/60 min = mm/hr.

Diffusion Across the Plasma Membrane

Some molecules can diffuse across a plasma membrane, but some cannot. In general, small, noncharged molecules can cross a membrane by simple diffusion, but large molecules cannot diffuse across a membrane. The dialysis tube membrane in the Experimental Procedure simulates a plasma membrane.

Experimental Procedure: Diffusion Across Plasma Membrane

At the start of the experiment,

1. Cut a piece of dialysis tubing approximately 40 cm (about 16 in) long. Soak the tubing in water until it is soft and pliable.
2. Close one end of the dialysis tubing with two knots.
3. Fill the bag halfway with glucose solution.
4. Add four full droppers of starch solution to the bag.

5. Hold the open end while you mix the contents of the dialysis bag. Rinse off the outside of the bag with distilled water.
6. Fill a beaker ⅔ full with distilled water.
7. Add droppers of iodine solution (IKI) to the water in the beaker until an amber (tealike) color is apparent.
8. Record the color of the solution in the beaker in Table 4.2.
9. Place the bag in the beaker with the open end hanging over the edge. Secure the open end of the bag to the beaker with a rubber band, as shown (Fig. 4.5). Make sure the contents do not spill into the beaker.

After about 5 minutes, at the end of the experiment,

10. You will note a color change. Record the color of the bag contents in Table 4.2.
11. Mark off a test tube at 1 cm and 3 cm.
12. Draw solution from near the bag and at the bottom of the beaker for testing with Benedict's reagent. Fill the test tube to the first mark with this solution. Add Benedict's reagent to the 3 cm mark. Heat in a boiling water bath for 5 to 10 minutes, observe any color change, and record your results as + or – in Table 4.2. (Optional use of glucose test strip: Dip glucose test strip into beaker. Compare stick with chart provided by instructor.)
13. Remove the dialysis bag from the beaker. Dispose of it and the used Benedict's reagent solution in the manner directed by your instructor.

Figure 4.5 Placement of dialysis bag in water containing iodine.

rubber band —
— open end of dialysis bag
dialysis membrane (simulates plasma membrane) —
— water and iodine solution
— glucose and starch
closed end of dialysis bag —

⚠ **Benedict's reagent** Exercise care in using this chemical. It is highly corrosive. If any should spill on your skin, wash the area with mild soap and water. Follow your instructor's directions for its disposal.

Table 4.2	Solute Diffusion Across Plasma Membrane				
At Start of Experiment			**At End of Experiment**		
	Contents	Color	Color	Benedict's Test	Conclusion
Bag	Glucose Starch			_____	
Beaker	Water Iodine				

Conclusions: Solute Diffusion Across the Plasma Membrane

- Based on the color change noted in the bag, conclude what solute diffused across the dialysis membrane from the beaker to the bag, and record your conclusion in Table 4.2.
- From the results of the Benedict's test on the beaker contents, conclude what solute diffused across the dialysis membrane from the bag to the beaker, and record your conclusion in Table 4.2.

- Which solute did not diffuse across the dialysis membrane from the bag to the beaker? _____

  How do you know? _____

  _____

  _____

4.3   Osmosis: Diffusion of Water Across Plasma Membrane

Osmosis is the diffusion of water across the plasma membrane of a cell. Just like any other molecule, water follows its concentration gradient and moves from the area of higher concentration to the area of lower concentration.

Experimental Procedure: Osmosis

To demonstrate osmosis, a thistle tube is covered with a membrane at its lower opening and partially filled with 50% corn syrup (starch solution) or a similar substance. The whole apparatus is placed in a beaker containing distilled water (Fig. 4.6). The water concentration in the beaker is 100%. Water molecules can move freely between the thistle tube and the beaker.

1. Note the level of liquid in the thistle tube, and measure how far it travels in 10 minutes:

    _____ mm

2. Calculate the speed of osmosis under these conditions: _____ mm/hr

Figure 4.6   Osmosis demonstration.
a. A thistle tube, covered at the broad end by a differentially permeable membrane, contains a corn syrup solution. The beaker contains distilled water. **b.** The solute is unable to pass through the membrane, but the water (arrows) passes through in both directions. There is a net movement of water toward the inside of the thistle tube, where there is a lower percentage of water molecules. **c.** Due to the incoming water molecules, the level of the solution rises in the thistle tube.

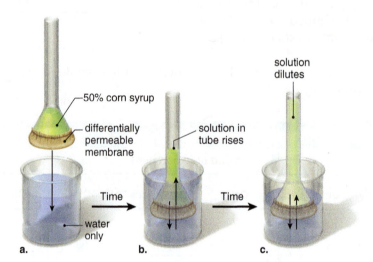

Conclusions: Osmosis

- In which direction was there a net movement of water? _____

    Explain what is meant by "net movement" after examining the arrows in Figure 4.6*b*.

    _____

- If the starch molecules in corn syrup moved from the thistle tube to the beaker, would there have

    been a net movement of water into the thistle tube? _____ Why wouldn't large starch molecules

    be able to move across the membrane from the thistle tube to the beaker? _____

    _____

- Explain why the water level in the thistle tube rose: In terms of solvent concentration, water moved from the area of _____ water concentration to the area of _____ water concentration across a differentially permeable membrane.

Tonicity in Cells

Tonicity is the relative concentration of solute (particles), and therefore also of solvent (water), outside the cell compared with inside the cell.

- An **isotonic solution** has the same concentration of solute (and therefore of water) as the cell. When cells are placed in an isotonic solution, there is no net movement of water (see Fig. 4.7*a*).
- A **hypertonic solution** has a higher solute (therefore, lower water) concentration than the cell. When cells are placed in a hypertonic solution, water moves out of the cell into the solution (see Fig. 4.7*b*).
- A **hypotonic solution** has a lower solute (therefore, higher water) concentration than the cell. When cells are placed in a hypotonic solution, water moves from the solution into the cell (see Fig.4.7*c*).

Animal Cells (Red Blood Cells)

A solution of 0.9% NaCl is isotonic to red blood cells (Fig. 4.7*a*). A solution greater than 0.9% NaCl is hypertonic to red blood cells. In such a solution, the cells shrivel up, a process called **crenation** (Fig. 4.7*b*). A solution of less than 0.9% NaCl is hypotonic to red blood cells. In such a solution, the cells swell to bursting, a process called **hemolysis** (Fig. 4.7*c*).

Figure 4.7   Tonicity and red blood cells.

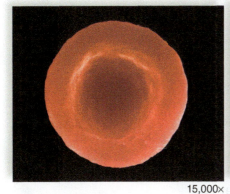
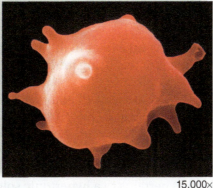
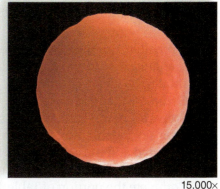

15,000×   15,000×   15,000×

a. Isotonic solution.
  Red blood cell has normal appearance due to no net gain or loss of water.

b. Hypertonic solution.
  Red blood cell shrivels due to loss of water.

c. Hypotonic solution.
  Red blood cell fills to bursting due to gain of water.

Experimental Procedure: Demonstration of Tonicity in Red Blood Cells

Three stoppered test tubes on display have the following contents:
  Tube 1: 0.9% NaCl plus a few drops of whole sheep blood
  Tube 2: 10% NaCl plus a few drops of whole sheep blood
  Tube 3: 0.9% NaCl plus distilled water and a few drops of whole sheep blood

 Do not remove the stoppers of test tubes during this procedure.

Laboratory 4   Cell Structure and Function   **39**

1. In the second column of Table 4.3, record the tonicity of each tube in relation to red blood cells.
2. Hold each tube in front of one of the pages of your lab manual. Determine whether you can see the print on the page through the tube. Record your findings in the fourth column of Table 4.3.
3. In the fifth column of Table 4.3, relate print visibility to effect of tonicity on the cells.

Table 4.3	Effect of Tonicity on Red Blood Cells			
Tube	Tonicity	Effect on Cells	Print Visibility	Explanation
1				
2				
3				

Plant Cells

When plant cells are in a hypotonic solution, such as fresh water, the large central vacuole gains water and exerts pressure, called **turgor pressure.** The cytoplasm, including the chloroplasts, is pushed up against the cell wall. You observed turgor pressure in Figure 4.3.

When plant cells are in a hypertonic solution, such as 10% NaCl, the central vacuole loses water, and the cytoplasm, including the chloroplasts, pulls away from the cell wall. This is called **plasmolysis.** You will observe plasmolysis in the following Experimental Procedure.

Experimental Procedure: Tonicity in Elodea Cells

1. If possible, use the *Elodea* slide you prepared earlier in this laboratory. If not, prepare a new wet mount of a small *Elodea* leaf using fresh water. Your slide should look like Figure 4.3.
2. Complete the portion of Table 4.4 that pertains to a hypotonic solution.
3. Prepare a new wet mount of a small *Elodea* leaf using a 10% NaCl solution.
4. After several minutes, focus on the surface of the cells. Note that plasmolysis has occurred and the cell contents are now in the center because the cytoplasm has pulled away from the cell wall due to loss of water from the large central vacuole.
5. Complete the portion of Table 4.4 that pertains to a hypertonic solution.

plasmolysis

Table 4.4	Effect of Tonicity on *Elodea* Cells	
Tonicity	Appearance of Cells	Due to (Scientific Term)
Hypotonic		
Hypertonic		

Experimental Procedure: Tonicity in Potato Strips

(This Experimental Procedure runs for one hour. Prior setup can maximize time efficiency.)

1. Cut two strips of potato, each about 7 cm long and 1.5 cm wide.
2. *Label two test tubes 1 and 2.* Place one potato strip in each tube.
3. Fill tube 1 with water to cover the potato strip.
4. Fill tube 2 with 10% sodium chloride (NaCl) to cover the potato strip.
5. After one hour, remove the potato strips from the test tubes and place them on a paper towel. Observe each strip for limpness (water loss) or stiffness (water gain). Which tube has the limp potato strip? _____ Use tonicity to explain why water diffused out of the potato strip in this tube. _____

   Which tube has the stiff potato strip? _____ Use tonicity to explain why water diffused into the potato strip in this tube. _____

   _____

6. Use this space to create a table to display your results. Give your table a title and columns for tube number and contents, tonicity, results, and explanation.

Conclusions: Tonicity

- In a hypotonic solution, animal cells _____. In red blood cells, this is called _____. In a hypertonic solution, animal cells _____. In red blood cells, this is called _____.

- In a hypotonic solution, the central vacuole of *Elodea* cells exerts _____ pressure, and chloroplasts are seen _____. In a hypertonic solution, the central vacuole loses water and _____ occurs. The cytoplasm plus the chloroplasts are seen _____.

- In a hypotonic solution, potato strips _____ water; in a hypertonic solution, potato strips _____ water and become _____.

4.4 pH and Cells

The pH of a solution tells its hydrogen ion concentration [H⁺]. The **pH scale** ranges from 0 to 14. A pH of 7 is neutral (Fig. 4.8). A pH lower than 7 indicates that the solution is acidic (has more hydrogen ions than hydroxide ions), whereas a pH greater than 7 indicates that the solution is basic (has more hydroxide ions than hydrogen ions). A **buffer** is a system of chemicals that takes up excess hydrogen ions or hydroxide ions, as appropriate.

The concept of pH is important in biology because organisms are very sensitive to hydrogen ion concentration. For example, in humans the pH of the blood must be maintained at about 7.4 or we become ill. All organisms need to maintain the hydrogen ion concentration, or pH, at a constant level.

Why are cells and organisms buffered? _____

_____

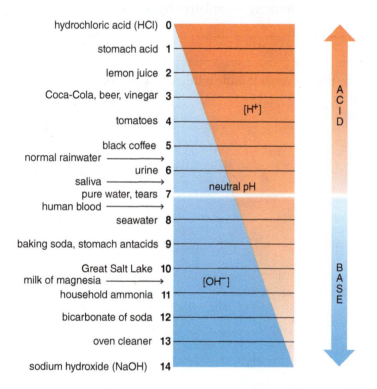

Figure 4.8   The pH scale.
The proportionate amount of hydrogen ions (H⁺) to hydroxide ions (OH⁻) is indicated by the diagonal line.

Experimental Procedure: pH and Cells

1. Label three test tubes, and fill them to the halfway mark as follows: tube 1: water; tube 2: buffer (inorganic) solution; and tube 3: simulated cytoplasm (buffered protein solution).

2. Use pH paper to determine the pH of each tube. Dip the end of a stirring rod into the solution, and then touch the stirring rod to a 5 cm strip of pH paper. Read the current pH by matching the color observed with the color code on the pH paper package. Record your results in the "pH Before Acid" column in Table 4.5.

3. Add 0.1 N hydrochloric acid (HCl) dropwise to each tube until you have added five drops—shake or swirl after each drop. Use pH paper as in step 2 to determine the new pH of each solution. Record your results in the "pH After Acid" column in Table 4.5.

> ⚠ **Hydrochloric acid (HCl)** used to produce an acid pH is a strong, caustic acid. Exercise care in using this chemical. If any HCl spills on your skin, rinse immediately with clear water. Follow your instructor's directions for disposal of tubes that contain HCl.

Table 4.5 pH and Cells

Tube	Contents	pH Before Acid	pH After Acid	Explanation
1	Water			
2	Buffer			
3	Cytoplasm			

Conclusions: pH and Cells

- Enter your explanations in the last column of Table 4.5.
- Why would you expect cytoplasm to be as effective as the buffer in maintaining pH? _____

_____

Experimental Procedure: Effectiveness of Antacids

This procedure tests the ability of commercial antacid tablets such as Alka-Seltzer, Rolaids, or Tums to absorb excess H^+.

1. Use a mortar and pestle to grind up the amount of antacid that is listed as one dose.
2. For each antacid tested, use 100 ml of phenol red solution diluted to a faint pink to wash the antacid into a 250 ml beaker. Phenol red solution is a pH indicator that turns yellow in an acid and red in a base. Use a stirring rod to get the powder to dissolve.
3. Add and count the number of 0.1 N HCl drops it takes for the solution to turn light yellow.
4. Record your results in Table 4.6.

Table 4.6 Effectiveness of Antacids

Antacid	Drops of Acid Needed to Reach End Point	Evaluation
1		
2		
3		

Conclusions: Effectiveness of Antacids

- Participate with others in concluding which of the antacids tested neutralizes the most acid.

_____

- Did dosage in mg have any effect on the results? _____

_____

- Which of the substances on the label could be a buffer? _____

1. What characteristic do all eukaryotic cells have in common? _____

   _____

2. Which organelle digests macromolecules and cell parts? _____

3. Why would you predict that an animal cell, but not a plant cell, might burst when placed in a hypotonic

   solution? _____

   _____

4. Which of the cellular organelles would be included in the following categories?

   a. Membranous canals and vacuoles _____

   _____

   b. Energy-related organelles _____

5. How do you distinguish between rough endoplasmic reticulum and smooth endoplasmic reticulum?

   a. Structure _____

   b. Function _____

6. If a dialysis bag filled with water is placed in a molasses solution, what do you predict will happen to

   the weight of the bag over time? _____

   Why? _____

   _____

7. What is the relationship between plant cell structure and the ability of plants to stand upright?

   _____

8. The police are trying to determine if material removed from the scene of a crime was plant matter. What

   would you suggest they look for? _____

   _____

9. A test tube contains red blood cells and a salt solution. When the tube is held up to a page, you cannot see

   the print. With reference to a concentration of 0.9% sodium chloride (NaCl), how concentrated is the salt

   solution? _____

10. Predict the microscopic appearance of cells in the leaf tissue of a wilted plant. _____

    _____

Essentials of Biology Website

Instructors can find lab prep information and answers to all of the laboratory questions in the Laboratory Resource Guide. *Students* can practice their knowledge with quizzes, animations, flashcards, and much more.

www.mhhe.com/maderessentials4

McGraw-Hill Access Science Website

An online encyclopedia of science and technology that provides information, including videos, that can enhance the laboratory experience.

www.accessscience.com

LEARNSMART

Cell Anatomy

Diffusion

Osmosis

pH & Cells

5
Enzymes

Introduction

The cell carries out many chemical reactions. All the chemical reactions that occur in a cell are collectively called **metabolism.** A possible chemical reaction can be indicated like this:

$$A + B \longrightarrow C + D$$
$$\text{reactants} \qquad \text{products}$$

In all chemical reactions, the **reactants** are molecules that undergo a change, which results in the **products.** The arrow means "produces," as in A + B produces C + D. The number of reactants and products can vary; in the one you are studying today, a single reactant breaks down to two products. All the reactions that occur in a cell have an enzyme. **Enzymes** are organic catalysts that speed metabolic reactions. Because enzymes are specific and speed only one type of reaction, they are given names. In today's laboratory, you will be studying the action of the enzyme **catalase.** The reactants in an enzymatic chemical reaction are called **substrate(s)** (Fig. 5.1).

**Figure 5.1
Enzymatic
action.**
A reaction such as
this degradation
reaction occurs on
the surface of
an enzyme at
the active site.
The enzyme is
reusable.

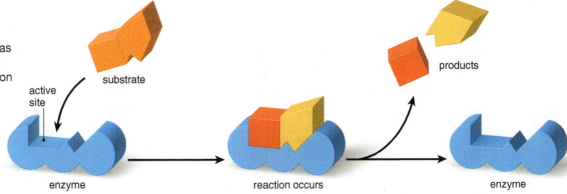

Enzymes are specific because they have a shape that accommodates the shape of their substrates. Enzymatic reactions can be indicated like this:

$$E + S \longrightarrow ES \longrightarrow E + P$$

In this reaction, E = enzyme, ES = enzyme-substrate complex, and P = product.

Two types of enzymatic reactions in cells are shown in Figure 5.1. During degradation reactions, the substrate is broken down to the product(s), and during synthesis reactions, the substrates are joined to form a product. A number of other types of reactions also occur in cells. The location where the enzyme and substrate form an enzyme-substrate complex is called the **active site** because the reaction occurs here. At the end of the reaction, the product is released, and the enzyme can then combine with its substrate again. A cell needs only a small amount of an enzyme because enzymes are used over and over. Some enzymes have turnover rates well in excess of a million product molecules per minute.

> 🔵 **Planning Ahead** To save time, your instructor may have you start a boiling water bath (page 48) at the beginning of the laboratory.

5.1   Catalase Activity

Catalase is involved in a degradation reaction: Catalase speeds the breakdown of hydrogen peroxide (H_2O_2) in nearly all organisms, including bacteria, plants, and animals. A cellular organelle called a peroxisome, which contains catalase, is present in every plant and animal organ. This means that we could use any plant or animal organ as our source of catalase today. Commonly, school laboratories use the potato as a source of catalase because potatoes are easily obtained and cut up.

Catalase performs a useful function in organisms because hydrogen peroxide is harmful to cells. Hydrogen peroxide is a powerful oxidizer that can attack and denature cellular molecules like DNA! Knowing its harmful nature, humans use hydrogen peroxide as a commercial antiseptic to kill germs (Fig. 5.2). In reduced concentration, hydrogen peroxide is a whitening agent used to bleach hair and teeth. Skillful technicians use it to provide oxygen to aquatic plants and fish, but it is also used industrially to clean most anything from tubs to sewage. It's even put in glow sticks, where it reacts with a dye that then emits light.

When catalase speeds the breakdown of hydrogen peroxide, water and oxygen are released:

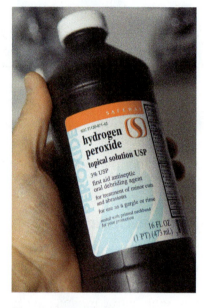

Figure 5.2   Hydrogen peroxide.
Bubbling occurs when you apply hydrogen peroxide to a cut because oxygen is being released when catalase, an enzyme present in the body's cells, degrades hydrogen peroxide.

$$\underset{\text{hydrogen peroxide}}{2\,H_2O_2} \xrightarrow{\text{catalase}} \underset{\text{water}}{2\,H_2O} + \underset{\text{oxygen}}{O_2}$$

What is the reactant in this reaction? _____ What is the substrate for

catalase? _____ What are the products in this reaction? _____ and

_____

Bubbling occurs as the reaction proceeds. Why? _____
In the Experimental Procedure that follows, you will use bubble height to indicate the amount of product per unit time and therefore enzyme activity. Examine Table 5.1 and hypothesize which tube (1, 2, or 3) will

have a greater bubble column height. Include a complete explanation in your hypothesis. _____

_____

_____

Label three clean test tubes (1 to 3), and use the appropriate graduated transfer pipet to add solutions to the test tubes as follows:

Tube 1
1. Add 1 ml of catalase buffered at pH 7.0, the optimum pH for catalase.
2. Add 4 ml of hydrogen peroxide. Swirl well to mix, and wait at least 20 seconds for bubbling to develop.
3. Measure the height of the bubble column (in mm), and record your results in Table 5.1.

Tube 2
1. Add 1 ml of water.
2. Add 4 ml of hydrogen peroxide. Swirl well to mix, and wait at least 20 seconds.
3. Measure the height of the bubble column and record your results in Table 5.1.

Tube 3
1. Add 1 ml of catalase.
2. Add 4 ml of sucrose solution. Swirl well to mix; wait 20 seconds.
3. Measure the height of the bubble column, and record your results in Table 5.1.

Table 5.1	Catalase Activity		
Tube	Contents	Bubble Column Height (mm)	Explanation
1	Catalase Hydrogen peroxide		
2	Water Hydrogen peroxide		
3	Catalase Sucrose solution		

Conclusions: Catalase Activity

- Which tube showed the amount of bubbling you expected? _____ Record your explanation in Table 5.1.

- Which tube is a negative control? _____ If this tube showed bubbling, what could you conclude about your procedure? _____
  Record your explanation in Table 5.1.

- Enzymes are specific; they speed only a reaction that contains their substrate. Which tube exemplifies this characteristic of an enzyme? _____ Record your explanation in Table 5.1.

5.2   Effect of Temperature on Enzyme Activity

The active sites of enzymes increase the likelihood that substrate molecules will find each other and interact. Therefore, enzymes lower the energy of activation (the temperature needed for a reaction to occur). Still, increasing the temperature is expected to increase the likelihood that active sites will be occupied because molecules move about more rapidly as the temperature rises. In this way, a warm temperature increases enzyme activity.

The shape of an enzyme and its active site must be maintained or else they will no longer be functional. A very high temperature, such as the one that causes water to boil, is likely to cause weak bonds of a protein to break; if this occurs, the enzyme **denatures**—it loses its original shape and the active site will no longer function to bring reactants together. Now enzyme activity plummets.

With this information in mind, examine Table 5.2 and hypothesize which tube (1, 2, or 3) will have more product per unit time as judged by bubble column height. Include a complete explanation in your hypothesis. _____

_____

Experimental Procedure: Effect of Temperature

Label three clean test tubes (1 to 3), and use the appropriate graduated transfer pipet to add solutions to the test tubes as follows:

1. To each tube add 1 ml of catalase buffered at pH 7.0, the optimum pH for catalase.
2. Place tube 1 in a refrigerator or cold water bath, tube 2 in an incubator or a warm water bath, and tube 3 in a boiling water bath. Complete the second column in Table 5.2. Wait 15 minutes.
3. As soon as you remove the tubes one at a time from the refrigerator, incubator, and boiling water, add 4 ml of hydrogen peroxide.
4. Swirl well to mix, and wait 20 seconds.
5. Measure the height of the bubble column in each tube, and record your results in Table 5.2. Plot your results in Figure 5.3.

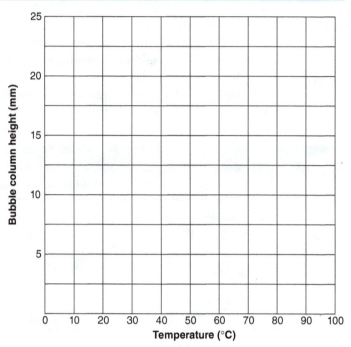

Figure 5.3   Effect of temperature on enzyme activity.

Table 5.2 Effect of Temperature			
Tube	Temperature °C	Bubble Column Height (mm)	Explanation
1 Refrigerator			
2 Incubator			
3 Boiling water			

Conclusions: Effect of Temperature

- The bubble column height indicates the degree of enzyme activity. Was your hypothesis supported? _____ Explain in Table 5.2 the degree of enzyme activity per tube.
- What is your conclusion concerning the effect of temperature on enzyme activity? _____

_____

5.3 Effect of Concentration on Enzyme Activity

Consider that if you increase the number of caretakers per number of children, it is more likely that each child will have quality time with a caretaker. So it is if you increase the amount of enzyme per amount of substrate—it is more likely that substrates will find the active site of an enzyme and a reaction will take place. With this in mind, examine Table 5.3 and hypothesize which tube (1, 2, or 3) will have more product per unit time as judged by bubble column height. Include a complete explanation in your

hypothesis. _____

_____

Experimental Procedure: Effect of Enzyme Concentration

Label three clean test tubes (1 to 3), and use the appropriate graduated transfer pipet to add solutions to the test tubes as follows:

Tube 1
1. Add 1 ml of water and 4 ml of hydrogen peroxide.
2. Swirl well to mix, and wait 20 seconds.
3. Measure the height of the bubble column, and record your results in Table 5.3.

Tube 2
1. Add 1 ml of buffered catalase and 4 ml of hydrogen peroxide.
2. Swirl well to mix, and wait 20 seconds.
3. Measure the height of the bubble column, and record your results in Table 5.3.

Tube 3
1. Add 3 ml of buffered catalase and 4 ml of hydrogen peroxide.
2. Swirl well to mix, and wait 20 seconds.
3. Measure the height of the bubble column, and record your results in Table 5.3.

Table 5.3	Effect of Enzyme Concentration		
Tube	**Amount of Enzyme**	**Bubble Column Height (mm)**	**Explanation**
1	none		
2	1 cm		
3	3 cm		

Conclusions: Effect of Concentration

- The bubble column height indicates the degree of enzyme activity. Was your hypothesis supported? _____ Explain in Table 5.3 the degree of enzyme activity per tube.
- If unlimited time were allotted, would the results be the same in all tubes? _____ Explain why or why not. _____
- Would you expect similar results if the substrate concentration were varied in the same manner as the enzyme concentration? _____ Why or why not? _____
- What is your conclusion concerning the effect of concentration on enzyme activity? _____
_____

5.4 Effect of pH on Enzyme Activity

Each enzyme has a pH at which the speed of the reaction is optimum (occurs best). Any higher or lower pH affects hydrogen bonding and the structure of the enzyme, leading to reduced activity.

⚠️ **Hydrochloric acid (HCl)** used to produce an acid pH is a strong, caustic acid, and sodium hydroxide (NaOH) used to produce a basic pH is a strong, caustic base. Exercise care in using these chemicals, and follow your instructor's directions for disposal of tubes that contain these chemicals. If any acidic or basic solutions spill on your skin, rinse immediately with clear water.

Catalase is an enzyme in cells where the pH is near 7 (called neutral pH). Other enzymes prefer different pHs. The pancreas secretes a slightly basic (below pH 7) juice into the digestive tract, and the stomach wall releases a very acidic digestive juice, which can be as low as pH 2. With this information about catalase in mind, examine Table 5.4 and hypothesize which tube (1, 2, or 3) will have more product per unit time as judged by the bubble column height. Include a complete explanation in your

hypothesis. _____

_____

Experimental Procedure: Effect of pH

Label three clean test tubes (1 to 3), and use the appropriate graduated transfer pipet to add solutions to the test tubes as follows:

Tube 1
1. Add 1 ml of nonbuffered catalase and 2 ml of water adjusted to pH 3 by the addition of HCl. Wait 1 minute.
2. Add 4 ml of hydrogen peroxide.
3. Swirl to mix, and wait 20 seconds.
4. Measure the height of the bubble column, and record your results in Table 5.4.

Tube 2
1. Add 1 ml of nonbuffered catalase and 2 ml of water adjusted to pH 7. Wait 1 minute.
2. Add 4 ml of hydrogen peroxide.
3. Swirl to mix, and wait 20 seconds.
4. Measure the height of the bubble column, and record your results in Table 5.4.

Tube 3
1. Add 1 ml of nonbuffered catalase and 2 ml of water adjusted to pH 11 by the addition of NaOH. Wait 1 minute.
2. Add 4 ml of hydrogen peroxide.
3. Swirl to mix, and wait 20 seconds.
4. Measure the height of the bubble column, and record your results in Table 5.4.

Plot your results from Table 5.4 here (Fig. 5.4).

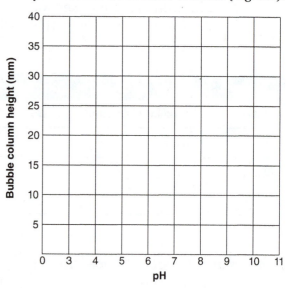

Figure 5.4   Effect of pH on enzyme activity.

Table 5.4	Effect of pH		
Tube	pH	Bubble Column Height (mm)	Explanation
1	3		
2	7		
3	11		

Conclusions: Effect of pH

- The amount of bubble column height indicates the degree of enzyme activity. Was your hypothesis supported? _____ Explain in Table 5.4 the degree of enzyme activity per tube.
- What is your conclusion concerning the effect of pH on enzyme activity? _____
  _____

5.5 Factors That Affect Enzyme Activity

In Table 5.5, summarize what you have learned about factors that affect the speed of an enzymatic reaction. For example, in general, what type of temperature promotes enzyme activity, and what type inhibits enzyme activity? Answer similarly for enzyme or substrate concentration and pH.

Table 5.5 Factors That Affect Enzyme Activity		
Factors	**Promote Enzyme Activity**	**Inhibit Enzyme Activity**
Enzyme specificity		
Temperature		
Enzyme or substrate concentration		
pH		

Conclusions: Factors That Affect Enzyme Activity

- Why does enzyme specificity promote enzyme activity? _____
  _____
- Why does a warm temperature promote enzyme activity? _____
  _____
- Why does increasing enzyme concentration promote enzyme activity? _____
  _____
- Why does optimum pH promote enzyme activity? _____
  _____

1. What happens at the active site of an enzyme? _____

2. On the basis of the active site, explain why the following conditions speed a chemical reaction:

   a. More enzyme _____

   b. More substrate _____

3. Name two other conditions (other than the ones mentioned in question 2) that maximize enzymatic reactions.

   a. _____

   b. _____

4. Explain the necessity for each of the two conditions you listed in question 3.

   a. _____

   b. _____

5. Lipase is a digestive enzyme that digests fat droplets in the basic conditions ($NaHCO_3$ is present) of the small intestine. Indicate which of the following test tubes would show digestion following incubation at 37°C, and explain why the others would not.

   Tube 1: Water, fat droplets _____

   Tube 2: Water, fat droplets, lipase _____

   Tube 3: Water, fat droplets, lipase, $NaHCO_3$ _____

   Tube 4: Water, lipase, $NaHCO_3$ _____

6. Fats are digested to fatty acids and glycerol. As the reaction described in question 5 proceeds, the

   solution will become what type pH? _____ Why? _____

   _____

7. Given the following reaction:

$$2\,H_2O_2 \xrightarrow{\text{catalase}} 2\,H_2O + O_2$$

$$\underset{\substack{\text{hydrogen}\\\text{peroxide}}}{} \qquad \underset{\substack{\text{water}\quad\text{oxygen}}}{}$$

   a. Which substance is the substrate? _____

   b. Which substance is the enzyme? _____

   c. Which substances are the end products? _____

   d. Is this a synthetic or degradative reaction? _____

   How do you know? _____

6

Photosynthesis

Introduction

The overall equation for **photosynthesis** is

$$CO_2 + H_2O \xrightarrow{\text{solar energy}} (CH_2O)_n + O_2$$

In this equation, (CH_2O) represents any general carbohydrate. Sometimes this equation is multiplied by 6 so that glucose $(C_6H_{12}O_6)$ appears as a product of photosynthesis. Photosynthesis takes place in chloroplasts (Fig. 6.1). Here membranous thylakoids are stacked in grana surrounded by the stroma. During the light reactions, pigments within the membranes, notably the chlorophylls, of thylakoids absorb solar energy, water (H_2O) is split, and oxygen (O_2) is released. The Calvin cycle reactions occur within the stroma. During these reactions, carbon dioxide (CO_2) is reduced and solar energy is now stored in a carbohydrate (CH_2O).

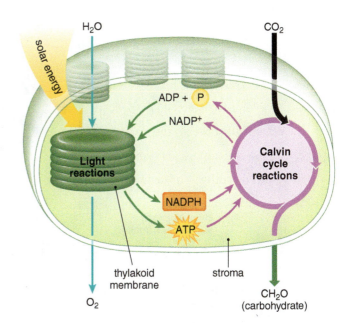

Figure 6.1 Overview of photosynthesis.

6.1 Plant Pigments

The principal pigment in the thylakoids of plants is **chlorophyll *a*. Chlorophyll *b*, carotenes,** and **xanthophylls** play a secondary role by transferring the energy they absorb to chlorophyll *a* for use in photosynthesis.

Chromatography is a technique that separates molecules from each other on the basis of their solubility in particular solvents. The solvents used in the following Experimental Procedure are petroleum ether and acetone, which have no charged groups and are therefore nonpolar. As a nonpolar solvent moves up the chromatography paper, the pigment moves along with it. The more nonpolar a pigment, the more soluble it is in a nonpolar solvent and the faster and farther it proceeds up the chromatography paper.

Experimental Procedure: Plant Pigments

1. Assemble a chromatography apparatus (large, dry test tube and cork with a hook) and a strip of precut chromatography paper (handle by the top only) (Fig. 6.2). Attach the paper strip to the hook and test for fit. The paper should hang straight and

> ⚠️ The chromatography solution is toxic and extremely flammable. Do not breathe the fumes, and do not place the chromatography solution near any source of heat. A fume (ventilation) hood is recommended.

barely touch the bottom of the test tube; trim if necessary. Measure 2 cm from the bottom of the paper and place a small dot with a pencil (not a pen). With a wax pencil, mark the test tube 1 cm below where the dot is with the stopper in place. Set the apparatus in a test tube rack.

Figure 6.2   Paper chromatography.
The paper must be cut to size and arranged to hang down without touching the sides of a dry tube. Then the pigment (chlorophyll) solution is applied to a designated spot. The chromatogram develops after the spotted paper is suspended in the chromatography solution.

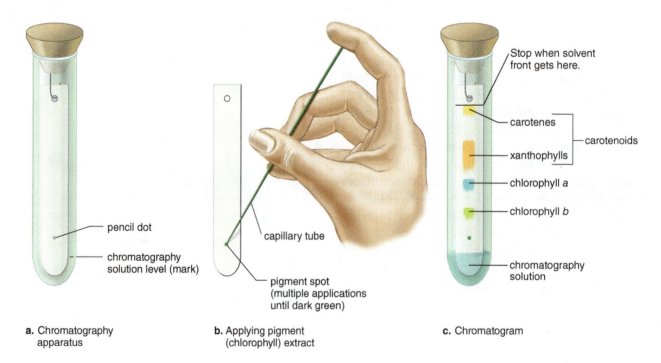

a. Chromatography apparatus

b. Applying pigment (chlorophyll) extract

c. Chromatogram

2. Prepare (or obtain) a plant *pigment extract,* as directed by your instructor.
3. Place the premarked chromatography paper strip onto a paper towel.
4. Fill a capillary tube by placing it into the extract. (It will fill by its own capillary action.)
5. Repeatedly apply the *pigment extract* to the pencil dot on the chromatographic strip. Let the spot dry between each application. Try to obtain a small, dark green spot. (Placing your index finger over the end of the capillary tube will help keep the dot small.)
6. In a **fume hood,** add *chromatography solution* to the mark you made earlier. Do not submerge the pigment spot. Set the apparatus in a test tube rack and close the chromatography apparatus tightly. Do not shake the test tube during the chromatography.
7. Allow approximately 10 minutes for your chromatogram to develop, but check it frequently so that the pigments do not reach the top of the paper.
8. When the solvent front has moved to within 1 cm of the upper edge of the paper (Fig. 6.2c), remove your chromatogram. Close the apparatus tightly. With a pencil, lightly mark the location of the solvent front, and allow the chromatogram to dry in the fume hood.
9. Identify the pigment bands. Carotenes are represented by the bright orange-yellow band at the top. Xanthophylls are yellow-brown and may be represented in multiple bands. The blue-green band is chlorophyll *a,* and the lowest, olive-green band is chlorophyll *b.* Which pigment is the

   most nonpolar (that is, has the greatest affinity for the nonpolar solvent)?_____
10. Calculate the R_f (ratio-factor) values for each pigment. For these calculations, mark the center of the initial pigment spot. This will be the starting point for all measurements. Also mark the midpoints of each pigment and the solvent front. Measure the distance between points for each pigment in millimeters (mm), and record these values in Table 6.1. Then use the following formula, and enter your R_f values in Table 6.1:

$$R_f = \frac{\text{distance moved by pigment}}{\text{distance moved by solvent}}$$

Table 6.1 Rf (Ratio-Factor) Values for Each Pigment		
Pigments	Distance Moved (mm)	R_f Values
Carotenes		
Xanthophylls		
Chlorophyll *a*		
Chlorophyll *b*		
Solvent		

11. Do your results suggest that the chemical characteristics of these pigments might differ?_____
    Why?_____

6.2 Solar Energy

During photosynthesis, solar energy is transformed into the chemical energy of a carbohydrate $(CH_2O)_n$. Without solar energy, photosynthesis would be impossible.

Release of oxygen from a plant indicates that photosynthesis is occurring. *Verify that photosynthesis releases oxygen by writing the equation for photosynthesis below:*

Role of White Light

White (sun) light contains different colors of light, as is demonstrated when white light passes through a prism (Fig. 6.3). White light is the best for photosynthesis because it contains all the colors of light. The oxygen released from photosynthesis is taken up by a plant when cellular respiration occurs. This must be taken into account when the rate of photosynthesis is calculated.

Figure 6.3   White light.
White light is made up of various colors, as can be seen when white light passes through a prism.

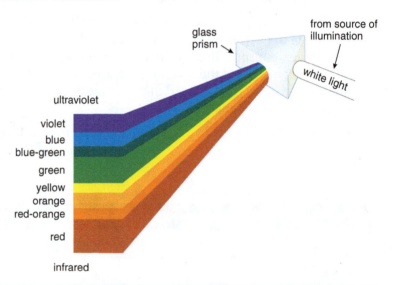

ultraviolet
violet
blue
blue-green
green
yellow
orange
red-orange
red
infrared

glass prism
from source of illumination
white light

Experimental Procedure: White Light

1. Place a generous quantity of *Elodea* with the cut end up (make sure the cuts are fresh) in a test tube with a rubber stopper containing a bent piece of glass tubing, as illustrated in Figure 6.4. When assembled, this is your volumeter for studying the need for light in photosynthesis. (Do not hold the volumeter in your hand, as body heat will also drive the reaction forward.) Your instructor will show you how to fix the volumeter in an upright position.
2. Before stoppering the test tube, add sufficient 3% *sodium bicarbonate* $(NaHCO_3)$ solution so that when the rubber stopper is inserted into the tube, the solution comes to rest at about 1/4 the length of the bent glass tubing. Mark this location on the glass tubing with a wax pencil.
3. Place a beaker of plain water next to the *Elodea* tube to serve as a heat absorber. Place a lamp (150 watt) next to the beaker. The tube, beaker, and lamp should be as close to one another as possible.
4. Turn on the lamp. As soon as the edge of the solution in the tubing begins to move, time the reaction for 10 minutes. Be careful not to bump the tubing or to readjust the stopper, or your readings will be altered. After 10 minutes, mark the edge of the solution, and measure in millimeters the distance

    the edge moved: _____ mm/10 min. This is **net photosynthesis,** a measurement that does not take into account the oxygen used up for cellular respiration.

    Record your results in Table 6.2. Why did the edge move forward? _____

Figure 6.4 Volumeter.
A volumeter apparatus is used to study the role of light in photosynthesis.

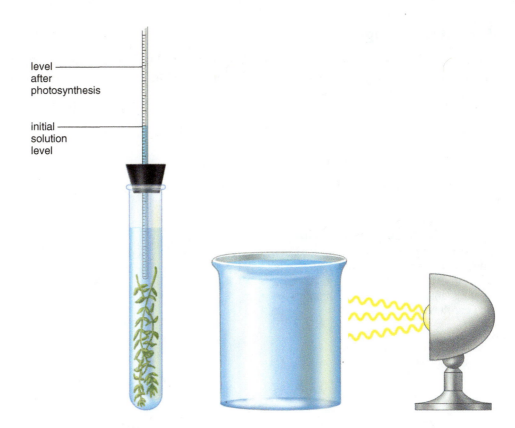

level after photosynthesis

initial solution level

5. Carefully wrap the tube containing *Elodea* in aluminum foil, and record here the length of time it takes for the edge of the solution in the tubing to recede 1 mm: _____. Convert your measurement to _____ mm/10 min, and record this value for **cellular respiration** in Table 6.2. (Do not use a minus sign, even though the edge receded.) Why does cellular respiration, which occurs in a plant all the time, cause the edge to recede? _____

_____

6. If the *Elodea* had not been respiring in step 4, how far would the edge have moved? _____mm/ 10 min. This is **gross photosynthesis** (net photosynthesis + cellular respiration). Record this number in Table 6.2.

7. Calculate the **rate of photosynthesis** (mm/hr) by multiplying gross photosynthesis (mm/10 min) by 6 (that is, 10 min × 6 = 60 min × 1 hr): _____ mm/hr. Record this value in Table 6.2.

Table 6.2 Rate of Photosynthesis (White Light)	
Data	
Net photosynthesis (white light)	
Cellular respiration (no light)	
Gross photosynthesis (net + cellular respiration)	(mm/10 min)
Rate of photosynthesis	(mm/hr)

Role of Green Light

Green light is only one part of white light (see Fig. 6.3). Plant pigments absorb certain colors of light better than other colors (Fig. 6.5). According to Figure 6.5, what color light do the chlorophylls

absorb best? _____ least? _____ What color light do the carotenoids (carotenes

and xanthophylls) absorb best? _____ least? _____

     Does photosynthesis use green light? _____

The following Experimental Procedure will test your answer.

Figure 6.5 Action spectrum for photosynthesis.
The action spectrum for photosynthesis is the sum of the absorption spectrums for the pigments chlorophyll *a*, chlorophyll *b*, and carotenoids.

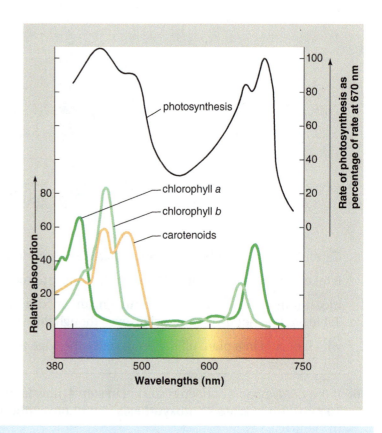

Experimental Procedure: Green Light

1. Add three drops of green dye (or use a green cellophane wrapper) to the beaker of water used in the previous Experimental Procedure until there is a distinctive green color. Remove all previous wax pencil marks from the glass tubing.
2. Record in Table 6.3 your data for gross photosynthesis (mm/10 min) and for rate of photosynthesis for white light (mm/hr) from Table 6.2.
3. Turn on the lamp. Mark the location of the edge of the solution on the glass tubing. As soon as the edge begins to move, time the reaction for 10 minutes. After 10 minutes, mark the edge of the solution, and measure in millimeters the distance the edge moved. Net photosynthesis

   for green light = _____ mm/10 min.
4. Carefully wrap the tube containing *Elodea* in aluminum foil, and record here the length of time

   it takes for the edge of the solution in the tubing to recede 1 mm: _____. Convert your

   measurement to _____ mm/10 min.

5. Calculate gross photosynthesis for green light (mm/10 min) as you did for white light, and record your data in Table 6.3.

6. Calculate rate of photosynthesis for green light (mm/hr) as you did for white light, and record your data in Table 6.3.

7. Average and record the Table 6.3 class data for both white light and green light, and record these averages in Table 6.3.

8. The following equation shows the rate of photosynthesis (green light) as a percentage of the rate of photosynthesis (white light):

$$\text{percentage} = \frac{\text{rate of photosynthesis (green light)}}{\text{rate of photosynthesis (white light)}} \times 100$$

This percentage, based on your data in Table 6.3, = _____. This percentage, based on class data in Table 6.3, = _____. Record these values in Table 6.3.

Table 6.3 Rate of Photosynthesis (Green Light)		
	Your Data	Class Data
Gross photosynthesis (mm/10 min)		
White (from Table 6.2)		
Green		
Rate of photosynthesis (mm/hr)		
White (from Table 6.2)		
Green		

Conclusions: Rate of Photosynthesis

- Explain why the rate of photosynthesis with green light is only a portion of the rate of photosynthesis with white light. _____

  _____

- How does the percentage based on your data differ from that based on class data?

  _____

  _____

6.3 Carbon Dioxide Uptake

During the second stage of photosynthesis, the plant takes up carbon dioxide (CO_2) and reduces it to a carbohydrate, such as glucose ($C_6H_{12}O_6$). Therefore, the carbon dioxide in the solution surrounding *Elodea* should disappear as photosynthesis takes place.

Experimental Procedure: Carbon Dioxide Uptake

1. Temporarily remove the *Elodea* from the test tube. Empty the sodium bicarbonate ($NaHCO_3$) solution from the test tube, rinse the test tube thoroughly, and fill it with a phenol red solution diluted to a faint pink. (Add more water if the solution is too dark.) Phenol red is a pH indicator that turns yellow in an acid and red in a base.
2. Blow *lightly* on the surface of the solution. Stop blowing as soon as the surface color changes to yellow. Then shake the test tube until the rest of the solution turns yellow.

    Blowing onto the solution adds what gas to the test tube? _____ When carbon dioxide combines with water, it forms carbonic acid. What causes the color change?

    _____

3. Thoroughly rinse the *Elodea* with distilled water, return it to the test tube with the phenol red solution, and assemble your volumeter as before.
4. If you used green dye, change the water in the beaker to remove the green solution.
5. Turn on the lamp and wait until the edge of the solution just begins to move. Note the time. Observe until you note a change in color. Record your results in the appropriate column of Table 6.4.

6. Hypothesize why the solution in the test tube eventually turned red. _____

    _____

Use of a Control

Scientists are more confident of their results when an experimental procedure includes a control. Controls undergo all the steps in the experiment except the one being tested.

- Considering the test sample in Table 6.4, suggest a possible control sample for this experiment:

    _____

- Ask your instructor if you can perform this procedure. Both the control and the test sample should be done at the same time.

- Record your results in Table 6.4. Why should all experiments have a control? _____

    _____

Table 6.4 Carbon Dioxide Uptake	
Tube	**Time for Color Change**
Test sample: *Elodea* + phenol red solution + CO_2	
Control sample:	

6.4 Carbon Cycle

In this laboratory, you have demonstrated a relationship between cellular respiration and photosynthesis. Animals produce the carbon dioxide plants use to carry out photosynthesis. Plants produce the food and oxygen that they and animals require to carry out cellular respiration. This relationship is illustrated in Figure 6.6 and can be represented by the following equation:

$$C_6H_{12}O_6 + 6\,O_2 \underset{\text{photosynthesis}}{\overset{\substack{\text{cellular} \\ \text{respiration}}}{\rightleftharpoons}} 6\,CO_2 + 6\,H_2O + \text{energy}$$

1. Which organelle in plants carries out the reaction in the previous equation in the reverse (right-to-left) direction? _____

2. Pertaining to photosynthesis, the energy in the equation is provided by _____.

3. Which organelle in plants and animals is involved in carrying out the reaction in this equation in the forward direction? _____

4. Pertaining to cellular respiration, the energy in the equation becomes chemical bond energy in what molecule? _____

5. Would it be correct to say that solar energy eventually becomes the chemical bond energy in ATP? _____ Why? _____

   _____

6. Considering that both plants and animals carry on cellular respiration, revise Figure 6.6 to improve its accuracy. Explain your changes.

   _____

   _____

Figure 6.6 Photosynthesis and cellular respiration.
Animals are dependent on plants for a supply of oxygen, and plants are dependent on animals for a supply of carbon dioxide.

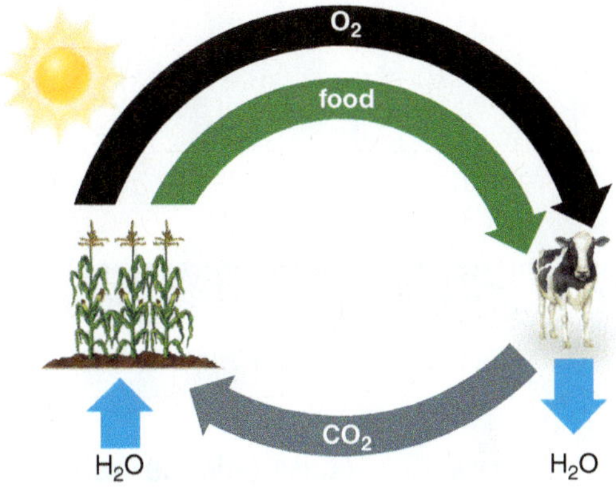

1. Where do the light reactions of photosynthesis take place? _____

2. What procedure did you use to separate plant pigments? _____

3. What determines the distance a pigment moves up the paper? _____
   _____

4. Where do plants ordinarily get the energy they need to carry on photosynthesis?
   _____

5. Blue, red, and green light are all present in what color of light? _____

6. Why do blue and red light, but not green, promote photosynthesis? _____
   _____

7. Does *Elodea* respire in the light or in the dark? Explain your answer.
   _____
   _____

8. Phenol red turns what color when carbon dioxide is added? _____

9. What happens to carbon dioxide during photosynthesis? _____
   _____

10. Some plants are colorless. Do you predict that they carry on photosynthesis? Explain. _____
    _____

Essentials of Biology Website

Instructors can find lab prep information and answers to all of the laboratory questions in the Laboratory Resource Guide. *Students* can practice their knowledge with quizzes, animations, flashcards, and much more.

www.mhhe.com/maderessentials4

McGraw-Hill Access Science Website

An online encyclopedia of science and technology that provides information, including videos, that can enhance the laboratory experience.

www.accessscience.com

Photosynthesis

7

Cellular Reproduction

Learning Outcomes

7.1 The Cell Cycle
- Name and describe the stages of the cell cycle, with emphasis on how previous stages prepare the cell for mitosis and cytokinesis.
- Relate a continual cell cycle to the development of cancer.
- Explain how DNA replication allows the chromosome number to stay constant during mitosis.

7.2 Animal Cell Mitosis and Cytokinesis
- Explain the importance of the mitotic spindle to mitosis.
- Identify the phases of mitosis in animal cell models and in a prepared animal mitosis slide.
- Describe how the animal cell undergoes cytokinesis.

7.3 Plant Cell Mitosis and Cytokinesis
- Identify the phases of mitosis in plant cell models and in a prepared plant mitosis slide.
- Recognize that plant cells exhibit the same phases of mitosis as the animal cell.
- Use the frequency of mitotic phases in a root tip slide to determine the time span for each phase of mitosis.
- Describe how the plant cell undergoes cytokinesis.

Introduction

Dividing cells experience nuclear division, cytoplasmic division, and a period of time between divisions called interphase. During **interphase,** the nucleus appears normal, and the cell is performing its usual cellular functions. Also, the cell is increasing all of its components, including such organelles as the mitochondria, ribosomes, and centrioles (in animal cells). DNA replication (making an exact copy of the DNA) occurs toward the end of interphase. Then, during nuclear division, called **mitosis,** the new nuclei receive the same number of chromosomes as the parental nucleus. When the cytoplasm divides, a process called **cytokinesis,** two daughter cells are produced (Fig. 7.1).

In multicellular organisms, mitosis permits growth and repair of tissues. In eukaryotic, unicellular organisms, mitosis is a form of asexual reproduction. Sexually reproducing organisms utilize another form of nuclear division, called **meiosis.** In animals, meiosis is a part of gametogenesis, the production of gametes (sex cells). The gametes are sperm in male animals and eggs in female animals. Meiosis is the topic for Laboratory 8.

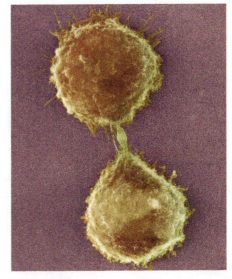

Figure 7.1   Cytokinesis.
Following chromosome distribution during mitosis, daughter cells are produced by cytokinesis, a division of the cytoplasm.

7.1   The Cell Cycle

As stated in the Introduction, the period of time between cell divisions is known as interphase. Because early investigators noted little visible activity between cell divisions, they dismissed this period as a resting state. But when later investigators discovered that DNA replication and chromosome duplication occur during interphase, the **cell cycle** concept was proposed. The cell cycle is divided into the four stages noted in Figure 7.2.

Figure 7.2   The cell cycle.
Immature cells go through a cycle consisting of four stages: G_1, S (synthesis), G_2, and M. The cell divides during the M stage, which consists of mitosis and cytokinesis.

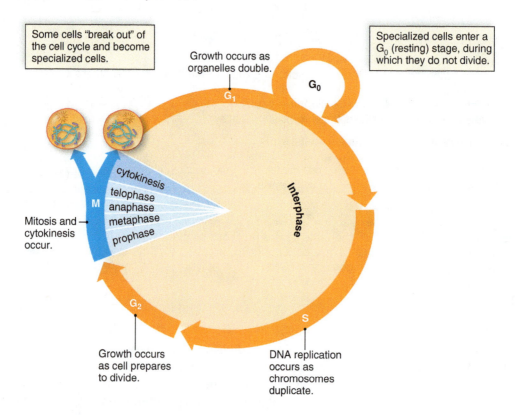

State the events of each stage of the cell cycle:

G_1 _____

S _____

G_2 _____

M _____

Explain why the entire process is called the "cell cycle." _____

The Cell Cycle and Cancer

Ordinarily, animal cells require about 18 to 24 hours for one cell cycle. Consult Figure 7.2, and notice that interphase consists of G_1, S, and G_2 stages of the cell cycle. When cancer occurs, cell cycle regulation is disturbed; interphase is severely shortened and very abnormal cells repeatedly undergo the cell cycle. A tumor develops, and treatment consists of shrinking or removing the tumor.

The S Stage of the Cell Cycle

Although each stage of the cell cycle is critical to successful mitosis (division of the nucleus), the S stage is particularly important because it is the stage during which DNA replicates. At the completion of replication, there are two double helices and each chromosome is duplicated.

When mitosis is about to occur, we can see each duplicated chromosome because chromatin has condensed and compacted to form two chromatids held together at a centromere. Consult Table 7.1, and *label the sister chromatids and centromere in Figure 7.3.*

What happens to the sister chromatids during mitosis? During mitosis, the sister chromatids separate, and then they are called daughter chromosomes. This is the manner in which DNA replication causes each body (somatic) cell of an organism to contain the same number of chromosomes. In humans, each cell in the body contains 46 chromosomes.

It is customary to call the cell that divides the **parent cell** and the resulting two cells the **daughter cells**. If a parent cell undergoing mitosis has 18 chromosomes, each daughter cell will have _____ chromosomes.

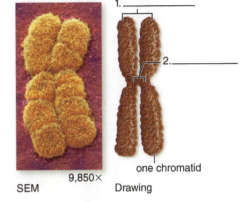

1. _____

2. _____

one chromatid

9,850×

SEM   Drawing

Figure 7.3 Duplicated chromosomes.
DNA replication results in duplicated chromosomes, which consists of two sister chromatids held together at a centromere.

The M Stage of the Cell Cycle

Study the terms in Table 7.1. These terms all pertain to mitosis. During mitosis sister chromatids (now called daughter chromosomes) move into the daughter nuclei. The process (mitosis) requires several phases. Consult Figure 7.2 and write the phases of mitosis here: _____

_____

Table 7.1	Structures Associated with Mitosis
Structure	**Description**
Nucleus	A large organelle containing the chromosomes and acting as a control center for the cells
Nucleolus	An organelle found inside the nucleus that produces the subunits of ribosomes
Chromosome	A rod-shaped body in the nucleus that is seen during mitosis and meiosis and that contains DNA, and therefore the hereditary units, or genes
Chromatids	The two identical parts of a chromosome following DNA replication
Centromere	A constriction where duplicates (sister chromatids) of a chromosome are held together
Spindle	A microtubule structure that brings about chromosome movement during cell division
Centrioles*	Short, cylindrical organelles at the spindle poles in animal cells
Aster*	Short, radiating fibers surrounding the centrioles in dividing cells

*Animal cells only

7.2 Animal Cell Mitosis and Cytokinesis

When an animal cell divides, it first undergoes mitosis, and then it undergoes cytokinesis, which is division of the cytoplasm. Mitosis is called duplication division because the daughter cells have the same chromosome makeup as the parent cell. As we now know, a spindle, sometimes called the mitotic spindle, occurs during mitosis. A **spindle** is composed of spindle fibers (microtubules) formed by the centrosomes, which are located at the **poles** of the spindle (Fig. 7.4). In animal cells, the poles contain centrioles surrounded by an **aster,** which is an array of fibers. The centromeres of the duplicated chromosomes are attached to spindle fibers at the equator of the spindle. When the centromeres split, the chromatids (now daughter chromosomes) move toward the poles.

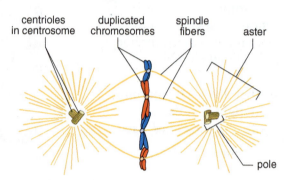

Figure 7.4   The mitotic spindle.
Each centromere of a duplicated chromosome is attached to spindle fibers at the equator of the spindle. When these spindle fibers shorten, the centromeres split and the sister chromatids, now called daughter chromosomes, move toward opposite spindle poles.

Animal Cell Mitosis

We will now have the opportunity to observe the stages of mitosis in animal cell models and a prepared slide. Mitosis is the type of nuclear division that occurs when embryos develop, organisms grow larger, and injuries heal. Without mitosis, none of these important events could occur.

Observation: Animal Cell Mitosis

Animal Mitosis Models

1. Each species has its own chromosome number. Counting the number of centromeres tells you the number of chromosomes in models or slides. What is the number of chromosomes in the parent cell

    and in the daughter cells in this model series? _____

2. Examine the phases of mitosis as depicted by the models.

3. Do these models show the *spindle,* which is illustrated in Figure 7.4? _____ In animal cells, the

    centrioles are surrounded by an *aster,* an array of fibers, at the poles of the spindle.

4. Name two ways you would be able to recognize animal cell mitosis: What is the shape of animal cells?

    _____

    What is the appearance of the spindle pole? _____

    Plant cells do not have centrioles; therefore, their spindle poles lack centrioles and their poles do not have asters.

Whitefish Blastula Slide

1. Examine a prepared slide of whitefish blastula cells undergoing mitosis. The blastula is an early embryonic stage in the development of animals.
2. Try to find a cell in each phase of mitosis using Figure 7.5 as a guide. Have a partner or your instructor check your ability to identify these cells.
3. Match each of the following statements to the correct phase of animal cell mitosis in Figure 7.5, and *write the correct statements on the lines provided.*

   Statements:
   - Daughter chromosomes are moving to the poles of the spindle.
   - Duplicated chromosomes have no particular arrangement in the cell.
   - Two daughter cells are now forming.
   - Duplicated chromosomes are aligned at the equator of the spindle.
4. The prophase cell in Figure 7.5 has the same number of chromosomes as the telophase nuclei. Explain the different appearance of the chromosomes. _____

_____

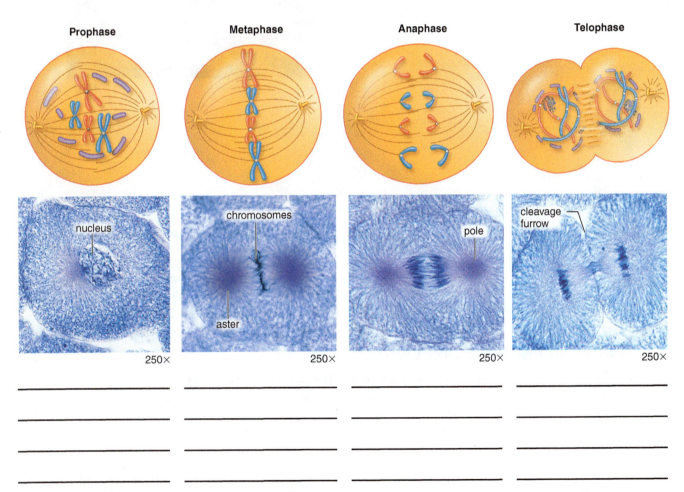

Figure 7.5   The phases of animal cell mitosis.
Mitosis always has these four main phases. Others, designated by the term "early" or "late," as in early prophase or late anaphase, can also be cited.

Cytokinesis in Animal Cells

Cytokinesis, division of the cytoplasm, usually accompanies mitosis. During cytokinesis, each daughter cell receives a share of the organelles that duplicated during interphase. Cytokinesis begins in anaphase, continues in telophase, and reaches completion by the start of the next interphase.

In animal cells, a **cleavage furrow,** an indentation of the membrane between the daughter nuclei, begins as anaphase draws to a close (Fig. 7.6). The cleavage furrow deepens as a band of actin filaments called the contractile ring slowly constricts the cell, forming two daughter cells. Are any of the cells in your whitefish blastula slide undergoing cytokinesis? _____

Do you see any cleavage furrows? _____

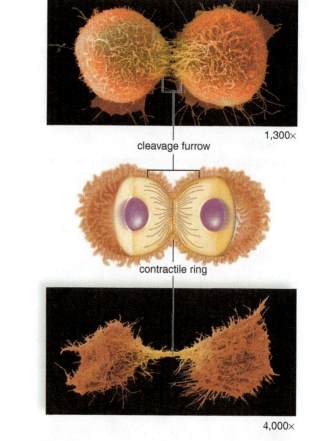

1,300×

cleavage furrow

contractile ring

4,000×

Figure 7.6 Cytokinesis in animal cells.
A single cell becomes two cells by a furrowing process. A contractile ring composed of actin filaments gradually gets smaller, and the cleavage furrow pinches the cell into two cells.

7.3   Plant Cell Mitosis and Cytokinesis

Division of a plant cell resembles that of an animal cell. The plant cell first undergoes mitosis; then it undergoes cytokinesis. However, plant cells, as you know, are surrounded by a plant cell wall. This feature has no effect on mitosis but it does affect cytokinesis, as we shall observe.

Plant Cell Mitosis

Although plant cells utilize a spindle to divide duplicated chromosomes, they do not have well-defined spindle poles because they lack centrioles and asters.

Interphase is not a phase *of* mitosis, but a preliminary stage *to* mitosis.

Plant Mitosis Models

1. Identify interphase and the phases of plant cell mitosis using models of plant cell mitosis and Figure 7.7 as a guide.
2. As mentioned previously, plant cells do not have centrioles and asters, which are short, radiating fibers produced by centrioles.
3. What is the number of chromosomes in each of the cells in your model series? _____

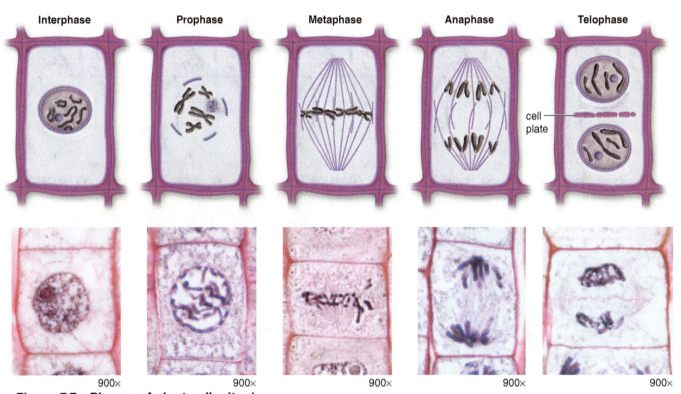

Figure 7.7   Phases of plant cell mitosis.
Mitosis has these phases in all eukaryotic cells. Plant cells are recognizable by the presence of a cell wall and the absence of centrioles and asters.

Onion Root Tip Slide

1. Examine a prepared slide of onion root tip cells *(Allium)* undergoing mitosis. In plants, the root tip contains tissue that is continually dividing and producing new cells (see Fig. 15.3).
2. Focus in low power and then switch to high power. Practice identifying phases that correspond to those shown in Figure 7.7.
3. Using high power, focus up and down on a cell in telophase. You may be able to just make out the cell plate, the region where a plasma membrane is forming between the two prospective daughter cells. Later, cell walls will appear in this region.

Time Span for Phases of the Cell Cycle in the Onion Root Tip

Knowing that the cell cycle consists of interphase plus four phases of mitosis and that the cell cycle typically lasts about 24 hours (1,440 minutes), hypothesize how many minutes the cell spends during each of these phases of the cell cycle.

Interphase _____ Prophase _____ Metaphase _____ Anaphase _____ Telophase _____

1. Select an area of the onion root tip slide that contains cells in all phases of mitosis and interphase. Concentrate on examining a confined area of about 20 to 30 cells.
2. Don't stray beyond a confined region, and as you identify phases, put a slash mark on the lines provided in this chart. Preferably work with a partner, who will enter the slash marks as you call out observed phases in a confined region of the root tip. Convert your 20 to 30 slashes to Arabic numbers on the lines provided under the column heading Total. Also record these numbers in the second column of Table 7.2.

Phase	Slash Marks	Total
Interphase	_____	_____
Prophase	_____	_____
Metaphase	_____	_____
Anaphase	_____	_____
Telophase	_____	_____

3. Calculate the percentage of total number of cells that are in each of the phases, and record the percentages in the third column of Table 7.2. (To do this, divide the number of cells in each phase by the total number of cells observed and multiply by 100.)
4. Assuming that the cell cycle lasts 24 hours (1,440 minutes), use these percentages to calculate the time span for each phase of the cell cycle. Enter the time span for each phase in the fourth column of Table 7.2.

Table 7.2 Time Span for Phases of the Cell Cycle in the Onion Root Tip			
Phase	Number Seen	% of Total	Time Span (min)
Interphase			
Prophase			
Metaphase			
Anaphase			
Telophase			
Total			

Conclusions: Time Span for Phases of the Cell Cycle in the Onion Root Tip

- Were your hypotheses supported or not supported by your observation of onion root tip cells undergoing the cell cycle? _____ Describe any specific discrepancies.
  _____
  _____

- Suggest a possible explanation for the length of time a cell spends on different phases of the cell cycle. _____
  _____

Cytokinesis in Plant Cells

After mitosis, the cytoplasm divides by cytokinesis. In plant cells, membrane vesicles derived from the Golgi apparatus migrate to the center of the cell and form a **cell plate** (Fig. 7.8), which is the location of a new plasma membrane for each daughter cell. Later, individual cell walls appear in this area. Were any of the cells of the onion root tip slide undergoing cytokinesis, as shown in Figure 7.7, during

Telophase? _____

How do you know? _____

Offer an explanation for why Figure 7.8 is so detailed. _____

_____

Would you predict that the vesicles of the cell plate lay down the new cell wall inside or outside the

vesicles? Explain your answer. _____

_____

Figure 7.8   Micrograph showing cytokinesis in plant cells.
During plant cell cytokinesis, vesicles fuse to form a cell plate that separates the daughter nuclei. Later, the cell plate gives rise to a new cell wall.

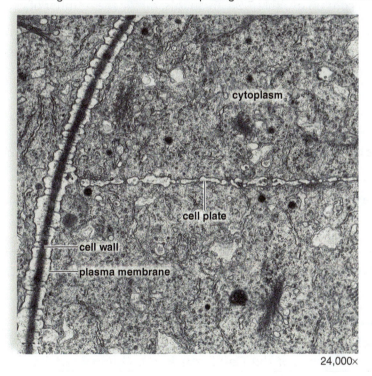

24,000×

Laboratory Review 7

1. Divide the cell cycle into two main portions, and tell in general what is happening in these two portions.

   _____

   _____

2. Most of the time the cell is in which of these portions of the cell cycle? Why is this advantageous? _____

   _____

   _____

3. Name and define the two events that take place when a cell divides. _____

   _____

   _____

4. What is the function of the centromere during mitosis? _____

   _____

5. Evolution of the spindle was of central importance to eukaryotic cells. What role is played by the spindle

   during mitosis? _____

   _____

6. Do the chromosomes have sister chromatids during these phases of mitosis (yes or no)?

   a. Prophase _____

   b. Metaphase _____

   c. Anaphase _____

   d. Telophase _____

7. Contrast the appearance of animal cells and plants during mitosis. _____

   _____

8. Explain how it is possible for each phase of mitosis and the daughter cells to have the same number of

   chromosomes. _____

   _____

9. Contrast how cytokinesis occurs in animal cells with how it occurs in plant cells. _____

   _____

   _____

10. What would a tissue look like if cytokinesis did not occur? _____

   _____

Essentials of Biology Website

Instructors can find lab prep information and answers to all of the laboratory questions in the Laboratory Resource Guide. *Students* can practice their knowledge with quizzes, animations, flashcards, and much more.

www.mhhe.com/maderessentials4

McGraw-Hill Access Science Website

An online encyclopedia of science and technology that provides information, including videos, that can enhance the laboratory experience.

www.accessscience.com

LEARNSMART

Mitosis & Meiosis

8

Sexual Reproduction

Introduction

In sexually reproducing organisms, meiosis is a part of or preparatory to gametogenesis, the production of gametes (sex cells). The gametes are the sperm (the smaller gamete) and egg (the larger gamete). Fusion of sperm and egg results in a zygote, which develops into a new individual (Fig. 8.1).

Meiosis is nuclear division that reduces the chromosome number so that the gametes have half the species number of chromosomes. At the start of meiosis, the parent cells have the full number of chromosomes and each is duplicated. Following two divisions, called **meiosis I** and **meiosis II,** the four daughter cells have only one copy of each type of chromosome, and these chromosomes consist of only one chromatid.

Not only does meiosis reduce the chromosome number, it also introduces variation, and this laboratory will show you exactly how the chromosomes are shuffled during meiosis and how the genetic material is recombined during the process of meiosis.

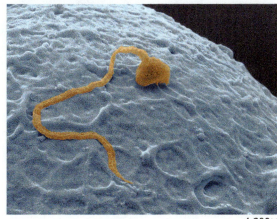

4,200×

Figure 8.1   Zygote formation.
During sexual reproduction, union of the sperm and egg produces a zygote, which becomes a new individual.

8.1 Meiosis: Reduction Division

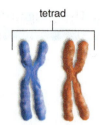

tetrad

Meiosis is a form of nuclear division that reduces the chromosome number by half. Therefore, when a sperm fertilizes an egg, the zygote will have the normal number of chromosomes for that species.

Before meiosis begins, the parent cell is **diploid (2n);** it contains pairs of chromosomes called **homologues.** Consider the pair of homologues shown in Figure 8.2. A pair of homologues is called a **tetrad** because it contains two pairs of sister chromatids, or four chromatids altogether. Following meiosis, the daughter cells are **haploid (n);** they contain only one from each pair of homologues.

Figure 8.2 Homologues.
A pair of homologues have the same shape and size. They form a tetrad because each homologue has two chromatids.

Before proceeding:

a. Distinguish between diploid (2n) and haploid (n): _____

_____

b. Distinguish between a homologue and a tetrad: _____

_____

Observation: Meiosis in Lily Anther

Almost all eukaryotes practice meiosis during sexual reproduction. It turns out that it's easier to examine meiosis in plants rather than animals because meiosis in plants is not a part of gametogenesis—that happens later. The anther is the male part of the flower where meiosis, consisting of **meiosis I** (Fig. 8.3) and **meiosis II** (Fig. 8.4), occurs preparatory to producing sperm.

Phases of Meiosis I

1. Examine a prepared slide of a lily anther where cells are undergoing meiosis I. Homologues pair up (called **synapsis**) during prophase I. During metaphase I, tetrads are at the equator and during anaphase I, homologues separate. Separation of homologues makes the daughter cells following meiosis I haploid because each daughter cell receives only one chromosome from each pair of homologues.

Figure 8.3 Meiosis I.
Phases of meiosis I in plant cell micrographs and drawings that depict the movement of the chromosomes.
(The blue chromosomes were inherited from one parent and the red chromosomes were inherited from the other parent.)

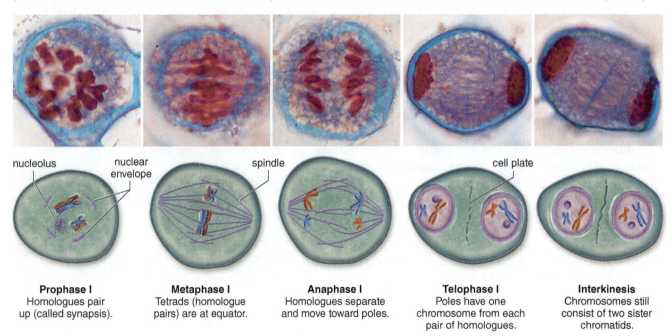

Prophase I	**Metaphase I**	**Anaphase I**	**Telophase I**	**Interkinesis**
Homologues pair up (called synapsis).	Tetrads (homologue pairs) are at equator.	Homologues separate and move toward poles.	Poles have one chromosome from each pair of homologues.	Chromosomes still consist of two sister chromatids.

nucleolus   nuclear envelope   spindle   cell plate

2. Using Figure 8.3 as a guide, try to find a cell in each phase of meiosis I. Tell what is happening in each of these phases:

   a. Prophase I _____

   b. Metaphase I _____

   c. Anaphase I _____

   d. Telophase I _____

3. *In Figure 8.3, place a 2n or an n beside each drawing.*

Phases of Meiosis II

1. Examine a prepared slide of a lily anther where cells are undergoing meiosis II. A brief period of time called **interkinesis** occurs between meiosis I and meiosis II. What happens during meiosis II? Separation of the sister chromatids occurs, of course, just as in mitosis (Fig. 8.4).

2. Using Figure 8.4 as a guide, try to find a cell in each phase of meiosis II. Tell what is happening in each of these phases:

   a. Prophase II _____

   b. Metaphase II _____

   c. Anaphase II _____

   d. Telophase II _____

3. *In Figure 8.4, place a 2n or an n beside each drawing.* Why is it important for both sperm and egg to have the haploid number of chromosomes? _____

   _____

Figure 8.4   Meiosis II.
Phases of meiosis II in plant cell micrographs and drawings that depict the movement of the chromosomes.

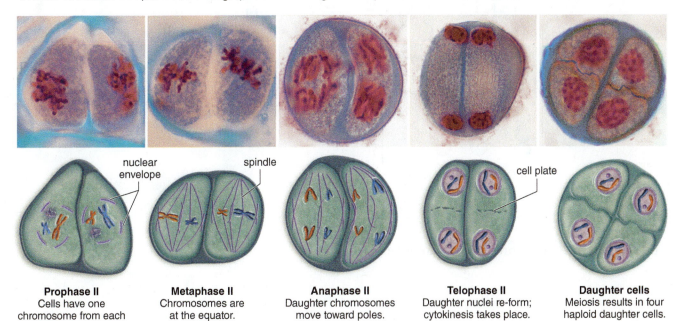

Prophase II
Cells have one chromosome from each pair of homologues.

Metaphase II
Chromosomes are at the equator.

Anaphase II
Daughter chromosomes move toward poles.

Telophase II
Daughter nuclei re-form; cytokinesis takes place.

Daughter cells
Meiosis results in four haploid daughter cells.

8.2 Production of Variation During Meiosis

The following Experimental Procedure is designed to show that, during meiosis, crossing-over and independent separation of homologues lead to diversity of genetic material in the gametes.

Experimental Procedure: Production of Variation During Meiosis

First, you will build four chromosomes: two pairs of homologues, as in Figure 8.5. In other words, the parent cell is 2n = 4.

Building Chromosomes

1. Obtain the following materials: 48 pop beads of one color (e.g., red) and 48 pop beads of another color (e.g., blue) for a total of 96 beads; eight magnetic centromeres.
2. Build a homologue pair of duplicated chromosomes using Figure 8.5a as a guide. Each chromatid will have 16 beads. Be sure to bring the centromeres of the same color together so that they form one duplicated chromosome.
3. Build another homologue pair of duplicated chromosomes, using Figure 8.5b as a guide. Each chromatid will have eight beads.
4. Note that your chromosomes look the same as those in Figure 8.5.

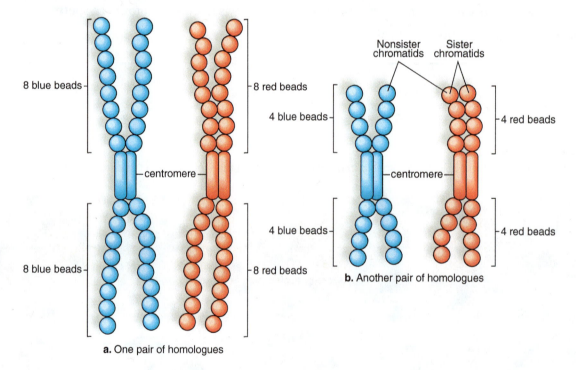

Figure 8.5   Two pairs of homologues.
The red chromosomes were inherited from one parent, and the blue chromosomes were inherited from the other parent. Color does not signify homologues; size and shape signify homologues.

Meiosis I

You will now have your chromosomes undergo meiosis I.

Prophase I

5. Put all four of your chromosomes in the center of your work area; this area represents the nucleus. Synapsis, a very important event, occurs during meiosis. To simulate synapsis, place the long blue chromosome next to the long red chromosome and the short blue chromosome next to the short red chromosome to show that the homologues pair up as in Figure 8.3 (prophase I). Now **crossing-over** occurs. During crossing-over an exchange of genetic material occurs between nonsister chromatids. Perform crossing-over by switching some blue beads for red beads between the inside chromatids of the homologues. The genetic material contains genes, and therefore crossing-over recombines genes.

Genetic Variation Due to Meiosis As a result of crossing-over, the genetic material on a chromosome in a gamete can be different from that in the parent cell.

Metaphase I

6. Keep the homologues together and align them at the equator. Add centrioles to represent the poles of a spindle apparatus. These are animal cells.

Anaphase I and Telophase I

7. Separate the homologues so that each pole of the spindle receives one chromosome from each pair of homologues. This separation causes each pole to receive the haploid number of chromosomes.

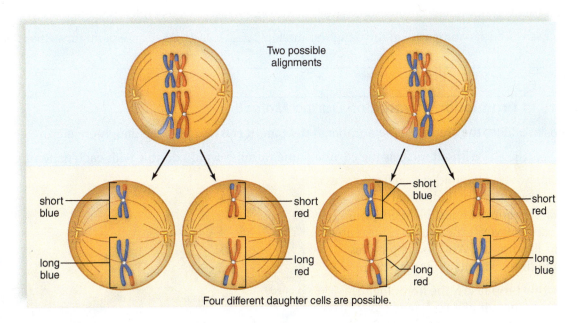

Four different daughter cells are possible.

Only two daughter cells result from meiosis I, but two possible alignments of chromosomes can occur at the equator during metaphase I. Therefore, four possible combinations of haploid chromosomes are possible in the daughter cells following meiosis I in a 2n = 4 parent cell. The possible combinations increase as the chromosome number increases.

Genetic Variation Due to Meiosis As a result of independent homologue separation, all possible combinations of the haploid number of chromosomes can occur among the gametes.

Meiosis II

Follow these directions to simulate meiosis II.

Prophase II

8. Choose one daughter nucleus (see step 7, page 77) to be the parent nucleus undergoing meiosis II.

Metaphase II

9. Move the duplicated chromosomes to the metaphase II equator, as shown in the art to the right.

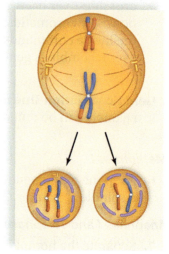

Anaphase II

10. Pull the two magnets of each duplicated chromosome apart. What does this action represent? _____

_____

Telophase II

11. Put the chromosomes—each having one chromatid—at the poles near the centrioles. At the end of telophase, the daughter nuclei re-form.
   - You chose only one daughter nucleus from meiosis I to be the nucleus that divides. In reality both daughter nuclei go on to divide again.

   Therefore, how many nuclei are usually present when meiosis II is complete? _____

   - In this exercise, how many chromosomes were in the parent cell nucleus undergoing meiosis II? _____

   - How many chromosomes are in the daughter nuclei? _____ Explain how this is possible.

_____

Summary of Production of Variation During Meiosis

1. Meiosis reduces the chromosome number. If the parent cell is 2n = 4, the daughter cells are

   n = _____. Without meiosis the chromosome number would double with each generation.

   Instead, when a haploid sperm fertilizes a haploid egg, the new individual is 2n.

2. Sexual reproduction results in offspring that can look very different, as represented in this illustration. Why do the puppies born to these parents show variation?

   a. During prophase I, the homologues come together and exchange genetic material. Now the inherited chromosomes will be different from those in the parent cell. This process is

   called _____.

   b. During metaphase I, the homologues align independently and therefore differently. This means that

   daughter cells following telophase I can have different _____ of chromosomes.

   c. During fertilization variant sperm fertilize variant eggs, further helping to ensure that the new

   individual inherits different _____ of homologues than a parent had.

8.3   Human Life Cycle

The term **life cycle** in sexually reproducing organisms refers to all the reproductive events that occur from one generation to the next. The human life cycle involves both mitosis and meiosis (Fig. 8.6). As you read the following text, *fill in boxes in Figure 8.6 with the terms "mitosis" or "meiosis."*

During development and after birth, mitosis is involved in the continued growth of the child and the repair of tissues at any time. As a result of mitosis, each somatic (body) cell has the diploid number of chromosomes (2n), which is 46 chromosomes.

During gamete formation, meiosis reduces the chromosome number from the diploid to the haploid number (n) in such a way that the gametes (sperm and egg) have one chromosome derived from each pair of homologues. In males, meiosis is a part of **spermatogenesis,** which occurs in the testes and produces sperm. In females, meiosis is a part of **oogenesis,** which occurs in the ovaries and produces eggs. After the sperm and egg join during fertilization, the resulting zygote is 2n. The zygote then undergoes mitosis with differentiation of cells to become a fetus, and eventually a new human being.

Meiosis keeps the number of chromosomes constant between the generations and, as we have seen, causes the gametes to be different from one another. Therefore, due to sexual reproduction, there are more variations among individuals.

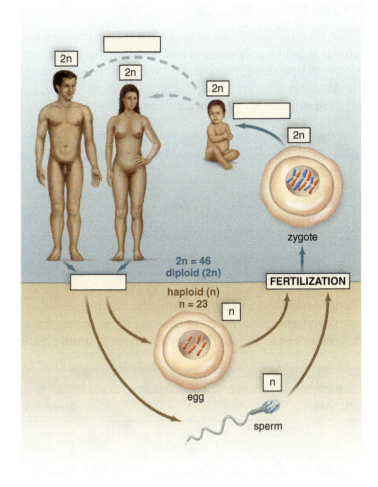

Figure 8.6   Life cycle of humans.

Summary of Human Life Cycle

Fill in the blanks to ensure your understanding of the role of meiosis and mitosis in humans.

1. Name of organ that produces gametes   in males _____   in females _____
2. Name of process that produces gametes   in males _____   in females _____
3. Type of cell division involved in process   in males _____   in females _____
4. Name of gamete   in males _____   in females _____
5. Number of chromosomes in gamete   in males _____   in females _____
6. Results of fertilization   _____
7. Number of chromosomes provided   _____

Mitosis Versus Meiosis

When comparing mitosis to meiosis it is important to note that meiosis requires two nuclear divisions but mitosis requires only one nuclear division. Therefore, mitosis produces two daughter cells and meiosis produces four daughter cells. Following mitosis, the daughter cells are still diploid, but following meiosis, the daughter cells are haploid. Figure 8.7 explains why.

Fill in Table 8.1 to indicate general differences between mitosis and meiosis.

Table 8.1 Differences Between Mitosis and Meiosis	Mitosis	Meiosis
1. Number of divisions		
2. Chromosome number in daughter cells		
3. Number of daughter cells		

Complete Table 8.2 to indicate specific differences between mitosis and meiosis I. Mitosis need be compared only with meiosis I because the same events occur during both mitosis and meiosis II, except that the cells are diploid during mitosis and haploid during meiosis II.

Table 8.2 Mitosis Compared with Meiosis I	
Mitosis	**Meiosis I**
Prophase: no pairing of chromosomes	Prophase I: _____
Metaphase: duplicated chromosomes at equator	Metaphase I: _____
Anaphase: sister chromatids separate	Anaphase I: _____
Telophase: chromosomes have one chromatid	Telophase I: _____

Provide the correct term for each definition:

1. Type of cell division that keeps the chromosome number the same and occurs during growth and repair _____

2. Type of cell division that reduces the chromosome number and occurs during gamete formation _____

3. Half the diploid number of chromosomes _____

4. Male gamete with the n number of chromosomes _____

5. Female gamete with the n number of chromosomes _____

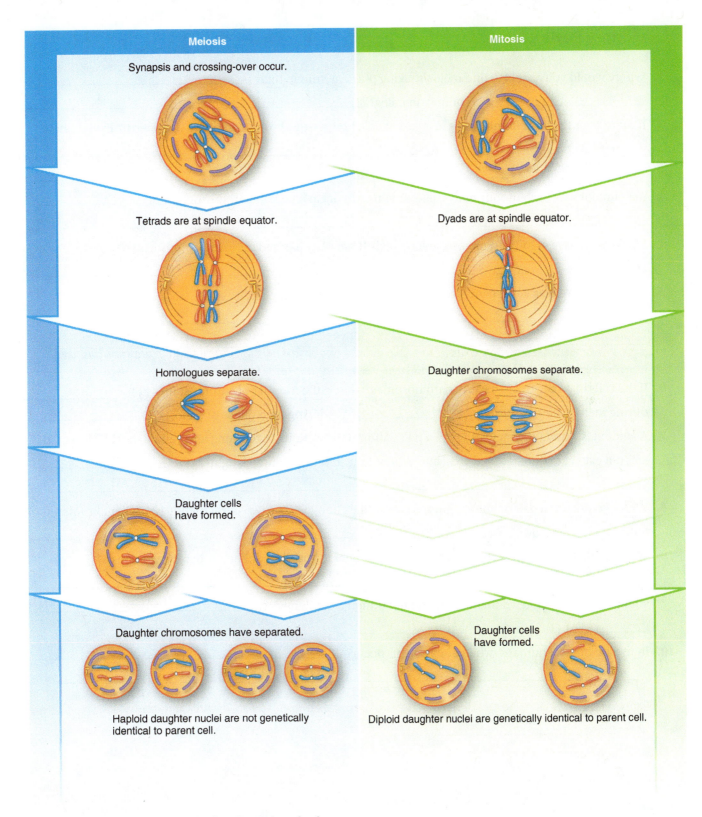

Meiosis	Mitosis
Synapsis and crossing-over occur.	
Tetrads are at spindle equator.	Dyads are at spindle equator.
Homologues separate.	Daughter chromosomes separate.
Daughter cells have formed.	
Daughter chromosomes have separated.	Daughter cells have formed.
Haploid daughter nuclei are not genetically identical to parent cell.	Diploid daughter nuclei are genetically identical to parent cell.

Figure 8.7   Comparison of mitosis and meiosis.
The blue chromosomes were inherited from one parent, and the red chromosomes were inherited from the other parent.

1. Where would you expect to find meiosis taking place in human males? _____ _____ in human females? _____

2. If there are 13 pairs of homologues in a parent cell, how many chromosomes are in a sperm? Explain how you arrived at this number. _____ _____

3. Account for why each of your body cells contains two of each kind of chromosome. _____ _____

4. Does metaphase of mitosis, meiosis I, or meiosis II have the haploid number of chromosomes at the equator of the spindle? _____

5. List four differences when comparing mitosis with meiosis. _____ _____ _____ _____

6. If the cells of an organism have 12 chromosomes, what is the number of chromosomes at the equator during metaphase of mitosis? _____ during metaphase of meiosis II? _____

7. A student is simulating meiosis I with a pair of homologues that are red-long and yellow-long. Why would you not expect to find both red-long and yellow-long in one resulting daughter cell? _____ _____

8. With reference to a pair of homologues, describe the change in the two participating nonsister chromatids following crossing-over. _____

9. What would be the appearance of a cell that completed mitosis but did not undergo cytokinesis? _____ _____ with a cell that completed meiosis but did not undergo cytokinesis? _____ _____

10. In the life cycle of humans, when does mitosis occur? _____ _____

Essentials of Biology Website

Instructors can find lab prep information and answers to all of the laboratory questions in the Laboratory Resource Guide. *Students* can practice their knowledge with quizzes, animations, flashcards, and much more.

www.mhhe.com/maderessentials4

McGraw-Hill Access Science Website

An online encyclopedia of science and technology that provides information, including videos, that can enhance the laboratory experience.

www.accessscience.com

LEARNSMART
Mitosis & Meiosis

9

Patterns of Inheritance

Learning Outcomes

9.1 One-Trait Crosses
- State Mendel's law of segregation, and relate this law to laboratory exercises.
- Describe and predict the results of a one-trait cross in both tobacco seedlings and corn.

9.2 Two-Trait Crosses
- State Mendel's law of independent assortment, and relate this law to laboratory exercises.
- Explain and predict the results of a two-trait cross in corn plants and *Drosophila*.

9.3 X-Linked Crosses
- Explain and predict the results of an X-linked cross in *Drosophila*.

Introduction

Gregor Mendel, sometimes called the "father of genetics," formulated the basic laws of genetics examined in this laboratory. He determined that individuals have two alternate forms of a gene (now called **alleles**) for each trait in their body cells. Today, we know that alleles are on the chromosomes. An individual can be **homozygous dominant** (two dominant alleles, *GG*), **homozygous recessive** (two recessive alleles, *gg*), or **heterozygous** (one dominant and one recessive allele, *Gg*). **Genotype** refers to an individual's genes, while **phenotype** refers to an individual's appearance (Fig. 9.1). Homozygous dominant and heterozygous individuals show the dominant phenotype; homozygous recessive individuals show the recessive phenotype.

Figure 9.1 Genotype versus phenotype.
Only with homozygous recessive do you immediately know the genotype.

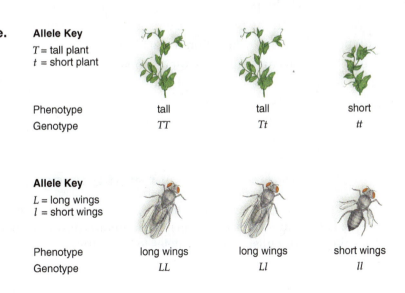

Allele Key
T = tall plant
t = short plant

Phenotype	tall	tall	short
Genotype	*TT*	*Tt*	*tt*

Allele Key
L = long wings
l = short wings

Phenotype	long wings	long wings	short wings
Genotype	*LL*	*Ll*	*ll*

Punnett Squares

Punnett squares, named after the man who first used them, allow you to easily determine the results of a cross between individuals whose genotypes are known. Consider that, when fertilization occurs, two gametes, such as a sperm and an egg, join together. Whereas individuals have two alleles for every trait, gametes have only one allele because alleles are on the chromosomes and homologues separate during meiosis. Heterozygous parents with the genotype *Aa* produce two types of gametes: 50% of the gametes contain an *A* and 50% contain an *a*. A Punnett square allows you to vertically line up all possible types of sperm and to horizontally line up all possible types of eggs. Every possible combination of gametes occurs within the squares, and these combinations indicate the genotypes of the offspring. In Figure 9.2, one offspring is *AA* = homozygous dominant, two are *Aa* = heterozygous, and one is *aa* = homozygous recessive. Therefore, three of the offspring will have the dominant phenotype and one individual will have the recessive phenotype. This is said to be a **phenotypic ratio** of 3:1.

A Punnett square can be used for any cross regardless of the trait(s) and the genotypes of the parents. All you need to do is use the correct letters for the particular trait(s), making sure you have given the parents the correct genotypes and correct proportion of each type gamete. Then you can determine the genotypes of the offspring and the resulting phenotypic ratio among the offspring.

Figure 9.2 What are the expected results of a cross?
A Punnett square allows you to determine the expected phenotypic ratio for a cross.

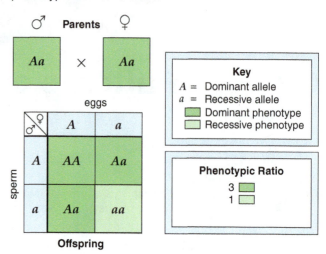

9.1 One-Trait Crosses

A single pair of alleles is involved in one-trait crosses. Mendel found that reproduction between two heterozygous individuals *(Aa)*, called a **monohybrid cross**, results in both dominant and recessive phenotypes among the offspring. In Figure 9.2, the expected phenotypic ratio among the offspring is 3:1. Three offspring have the dominant phenotype for every one that has the recessive phenotype.

Mendel realized that these results are obtainable only if the alleles of each parent segregate (separate from each other) during meiosis. Therefore, Mendel formulated his first law of inheritance:

Law of Segregation

Each organism contains two alleles for each trait, and the alleles segregate during the formation of gametes. Each gamete (egg or sperm) then contains only one allele for each trait. When fertilization occurs, the new organism has two alleles for each trait, one from each parent.

Inheritance is a game of chance. Just as there is a 50% probability of heads or tails when tossing a coin, there is a 50% probability that a sperm or an egg will have an *A* or an *a* when the parent is *Aa*. The chance of an equal number of heads or tails improves as the number of tosses increases. In the same way, the chance of an equal number of gametes with *A* and *a* in a cross improves as the number of gametes increases. Therefore, the 3:1 ratio among offspring is more likely the more offspring you count for the same type cross.

Color of Tobacco Seedlings

In tobacco plants, a dominant allele *(C)* for chlorophyll gives the plants a green color, and a recessive allele *(c)* for chlorophyll causes a plant to appear white. If a tobacco plant is homozygous for the recessive allele *(c)*, it cannot manufacture chlorophyll and thus appears white (Fig. 9.3).

Figure 9.3 Monohybrid cross.
These tobacco seedlings are growing on an agar plate. The white plants cannot manufacture chlorophyll.

Experimental Procedure: Color of Tobacco Seedlings

1. Obtain a numbered agar plate on which tobacco seedlings are growing. They are the offspring of a cross between heterozygous parents: the cross *Cc* × *Cc*. Complete the Punnett square to determine the expected phenotypic ratio.

   What is the expected phenotypic ratio? _____

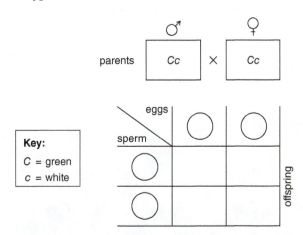

2. Record the plate number and, using a stereomicroscope, view the seedlings. Count the number that are green and the number that are white. Record your results in Table 9.1.
3. Repeat this procedure for two additional plates. Total the number that are green and the number that are white.
4. Complete Table 9.1 by recording the class data. Total the number that are green and the number that are white per class.

Table 9.1 Color of Tobacco Seedlings

	Number of Offspring		
	Green Color	White Color	Phenotypic Ratio
Plate # _____			
Plate # _____			
Plate # _____			
Totals			
Class data			

Conclusions: Color of Tobacco Seedlings

- In the last column of Table 9.1, record the actual phenotypic ratio per observed plate; per total number of green versus white plants you counted; and per the entire class. To determine the actual phenotypic ratio, divide the number of green seedlings in a plate by the number of white seedlings in a plate.

  Do your results differ from the expected phenotypic ratio? _____ If so, explain.

  _____

  _____

- Mendel found that the more plants he counted, the closer he came to the expected phenotypic ratio.

  Was your class data closer to the expected phenotypic ratio than your individual data? _____

  This is expected, because the more crosses you observe, the more likely it is that all types of sperm and eggs will have a chance to come together.

Color of Corn Kernels

In corn plants, the allele for purple kernel *(P)* is dominant over the allele for yellow kernel *(p)* (Fig. 9.4).

Figure 9.4 Monohybrid cross.
Two types of kernels are seen on an ear of corn following a monohybrid cross (Pp × Pp): purple and yellow.

1. Obtain an ear of corn from the supply table. You will be examining the results of the cross *Pp* × *pp*. Complete the Punnett square to determine the expected phenotypic ratio. Note that when one parent has only one possible type of gamete, only one column is needed in the Punnett square.

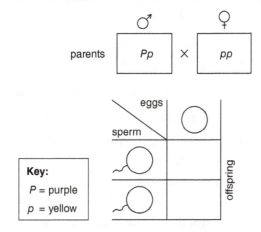

What is the expected phenotypic ratio among the offspring? _____

2. Record the sample number, and then count the number of kernels that are purple and the number that are yellow. As before, use two more samples, and record your results in Table 9.2.

Table 9.2	Color of Corn Kernels		
	Number of Kernels		
	Purple	**Yellow**	**Phenotypic Ratio**
Sample # _____			
Sample # _____			
Sample # _____			
Totals			
Class data			

Conclusions: Color of Corn Kernels

- In the last column of Table 9.2, record the actual phenotypic ratio per observed sample; per total number of purple versus yellow kernels you counted; and per the entire class. Do your results differ from the expected phenotypic ratio? _____ If so, explain. _____

  _____

- Was your class data closer to the expected phenotypic ratio than your individual data? _____ If so, explain. _____

  _____

1. In pea plants, purple flower color *(P)* is dominant and white flower color *(p)* is recessive. What is the genotype of pure-breeding white plants? "Pure-breeding" means that they produce plants with only one phenotype. _____ If pure-breeding purple plants are crossed with these white plants, what phenotype is expected? _____

2. In pea plants, tall *(T)* is dominant and short *(t)* is recessive. A heterozygous tall plant is crossed with a short plant. What is the expected phenotypic ratio? _____

3. Unexpectedly to the farmer, two tall plants have some short offspring. What are the genotypes of the parent plants and the short offspring? parent _____ offspring _____

4. In horses, two trotters are mated to each other and produce only trotters; two pacers are mated to each other and produce only pacers. When one of these trotters is mated to one of the pacers, all the horses are trotters. Create a key and show the cross. key _____ cross _____

5. A brown dog is crossed with two different black dogs. The first cross produces only black dogs, and the second cross produces equal numbers of black and brown dogs. What is the genotype of the brown dog? _____ the first black dog? _____ the second black dog? _____

6. In pea plants, green pods *(G)* are dominant and yellow pods *(g)* are recessive. When two pea plants with green pods are crossed, 25% of the offspring have yellow pods. What are the genotypes of all plants involved? plants with green pods? _____ plants with yellow pods? _____

7. A breeder wants to know if a dog is homozygous black or heterozygous black. If the dog is heterozygous, which cross is more likely to produce a brown dog, *Bb* × *bb* or *Bb* × *Bb?* Explain your answer.

   _____

   _____

8. If the cross in question 6 produces 220 plants, how many offspring have green pods and how many have yellow pods? _____ If the cross in question 2 produces 220 plants, how many offspring are tall and how many are short? _____

9.2 Two-Trait Crosses

Two-trait crosses involve two pairs of alleles. Mendel found that during a **dihybrid cross,** when two dihybrid individuals *(AaBb)* reproduce, the phenotypic ratio among the offspring is 9:3:3:1, representing four possible phenotypes. He realized that these results could be obtained only if the alleles of the parents segregated independently of one another when the gametes were formed. From this, Mendel formulated his second law of inheritance:

Law of Independent Assortment

Members of an allelic pair segregate (assort) independently of members of another allelic pair. Therefore, all possible combinations of alleles can occur in the gametes.

The FOIL method is a way to determine the gametes. FOIL stands for *First* two alleles from each trait; *Outer* two alleles from each trait; *Inner* two alleles from each trait; *Last* two alleles from each trait. Here is how the FOIL method can help you determine the gametes for the genotype *PpSs*:

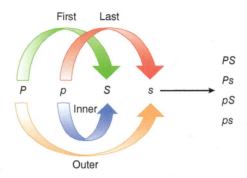

Color and Texture of Corn

In corn plants, the allele for purple kernel *(P)* is dominant over the allele for yellow kernel *(p),* and the allele for smooth kernel *(S)* is dominant over the allele for rough kernel *(s)* (Fig. 9.5).

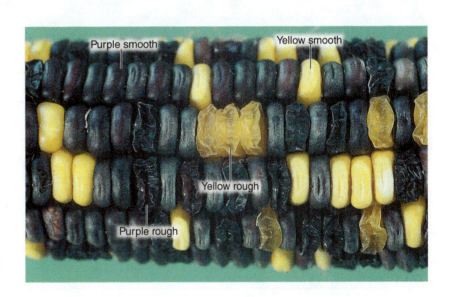

Figure 9.5 Dihybrid cross.
Four types of kernels are seen on an ear of corn following a dihybrid cross: purple smooth, purple rough, yellow smooth, and yellow rough.

Experimental Procedure: Color and Texture of Corn

1. Obtain an ear of corn from the supply table. You will be examining the results of the cross *PpSs* × *PpSs.*

2. Do the Punnett square on page 90 in order to state the expected phenotypic ratio among the offspring. _____

3. Count the number of kernels of each possible phenotype listed in Table 9.3. Record the sample number and your results in Table 9.3. Use three samples, and total your results for all samples. Also record the class data (i.e., the number of kernels that are the four phenotypes per class).

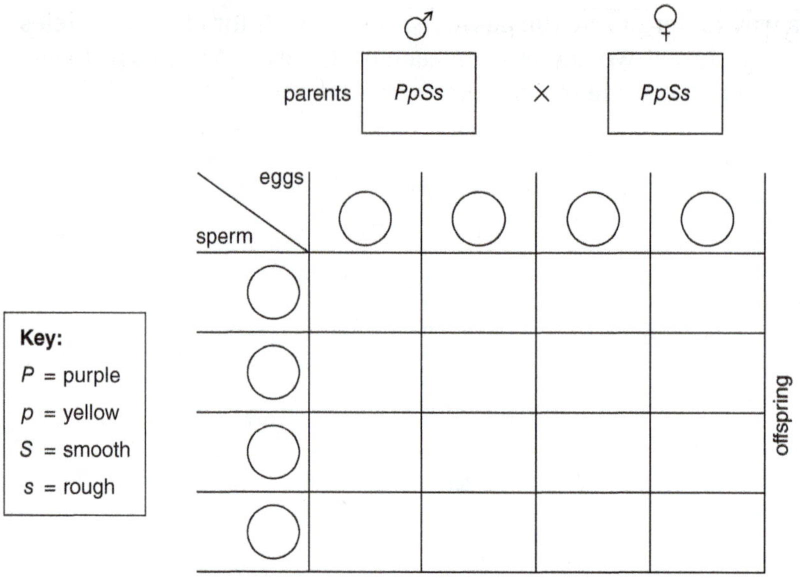

parents | $PpSs$ | × | $PpSs$

eggs

sperm

offspring

Key:

P = purple
p = yellow
S = smooth
s = rough

Table 9.3   Color and Texture of Corn

	Number of Kernels				
	Purple Smooth	Purple Rough	Yellow Smooth	Yellow Rough	Phenotypic Ratio
Sample # _____					
Sample # _____					
Sample # _____					
Totals					
Class data					

Conclusions: Color and Texture of Corn

- Calculate the actual phenotypic ratios based on the data, and record them in Table 9.3. Do the results differ from the expected ratio per your data? _____ per class data? _____ If so, explain.

_____

_____

_____

Wing Length and Body Color in *Drosophila*

Drosophila are the tiny flies you often see flying around ripe fruit; therefore, they are called fruit flies. If a culture bottle of fruit flies is on display, take a look at it. Because so many flies can be grown in a small culture bottle, fruit flies have contributed substantially to our knowledge of genetics. If you were to examine *Drosophila* flies under the stereomicroscope, they would appear like this:

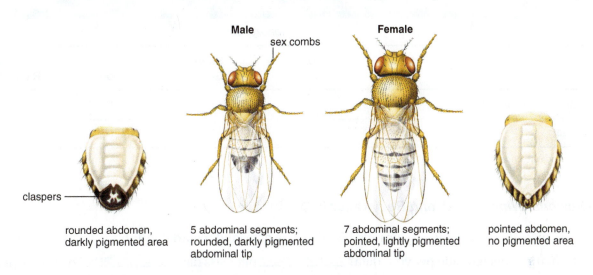

In *Drosophila*, long wings *(L)* are dominant over short (vestigial) wings *(l)*, and gray body *(G)* is dominant over ebony (black) body *(g)*. Consider the cross *LlGg* × *llgg* and complete this Punnett square:

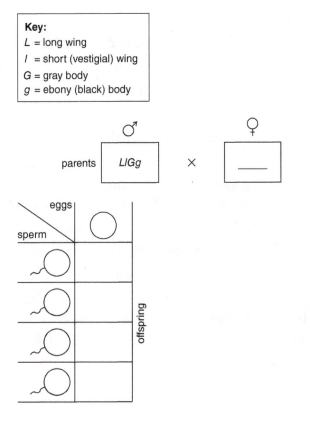

What is the expected phenotypic ratio for this cross? _____

If your instructor has frozen flies available, cross out the numbers in Table 9.4 and use the stereomicroscope or a hand lens to count the flies of each type given in Table 9.4. Otherwise, simply use the data supplied for you in Table 9.4.

Table 9.4   Wing Length and Body Color in *Drosophila**

	Phenotypes				
	Long Gray	Long Ebony	Short Gray	Short Ebony	Phenotypic Ratio
Number of offspring	28	32	28	30	
Class data	128	120	120	120	

*Wings and body are understood in this table.

Conclusions: Wing Length and Body Color in Drosophila

- Calculate the actual phenotypic ratio based on the data, and record it in Table 9.4. Do the results differ from the expected ratio per your data? _____ per class data? _____ If so, explain.

_____

_____

Two-Trait Genetics Problems

1. In tomatoes, tall is dominant and short is recessive. Red fruit is dominant and yellow fruit is recessive. Choose a key for height: _____ for color of fruit: _____ What is the genotype of a plant heterozygous for both traits? _____ What are the possible gametes for this plant?

_____

2. Using words, what are the likely parental genotypes if the results of a two-trait problem are 1:1:1:1 among the offspring? _____ × _____

3. In horses, black *(B)* and a trotting gait *(T)* are dominant, while brown *(b)* and a pacing gait *(t)* are recessive. If a black trotter (homozygous for both traits) is mated to a brown pacer, what phenotypic ratio is expected among the offspring? _____

4. Two black trotters have a brown pacer offspring. What are the genotypes of all horses involved? black trotter parents _____ brown pacer offspring _____

5. The phenotypic ratio among the offspring for two corn plants producing purple and smooth kernels is 9:3:3:1. (See page 90 for the key.) What is the genotype of these plants? parental plants _____ the 9 offspring _____ 3 of the offspring _____ the other 3 offspring _____ the 1 offspring _____

6. Which matings could produce at least some fruit flies heterozygous in both traits? Write yes or no beside each. (You do not need a key.)

ggLl × Ggll _____   GGLl × ggLl _____   GGLL × ggll _____

Explain your answers. _____

_____

7. State two new crosses that could not produce fruit flies heterozygous in both traits.

_____ × _____   _____ × _____

8. Chimpanzees are not deaf if they inherit both an allele *E* and an allele *G*. A cross between two deaf chimpanzees produces only chimpanzees that can hear. What are the genotypes of all chimpanzees involved? parents _____ × _____ offspring _____

9.3   X-Linked Crosses

In animals such as fruit flies, chromosomes differ between the sexes. All but one pair of chromosomes in males and females are the same; these are called **autosomes** because they do not actively determine sex. The pair that is different is called the **sex chromosomes.** In fruit flies and humans, the sex chromosomes in females are XX and those in males are XY.

Some alleles on the X chromosome have nothing to do with gender, and these genes are said to be X-linked. The Y chromosome does not carry these genes and indeed carries very few genes. Males with a normal chromosome inheritance are never heterozygous for X-linked alleles, and if they inherit a recessive X-linked allele it will be expressed.

Red/White Eye Color in *Drosophila*

In fruit flies, red eyes (X^R) are dominant over white eyes (X^r). You will be examining the results of the cross $X^R Y \times X^R X^r$. Complete this Punnett square and state the expected phenotypic ratio for this cross.

females _____   males _____

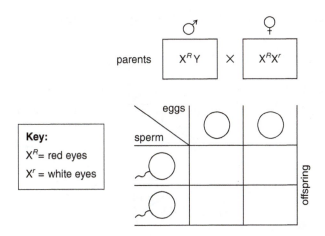

If your instructor has frozen flies available, cross out the numbers in Table 9.5 and use the stereomicroscope or a hand lens to count the flies of each type given in Table 9.5. Use the art on page 91 to tell males from females, and record male and female data separately. If frozen flies are not available, simply use the data supplied for you in Table 9.5.

Table 9.5   Red/White Eye Color in *Drosophila*

	Number of Offspring		
Your Data	**Red Eyes**	**White Eyes**	**Phenotypic Ratio**
Males	16	17	
Females	63	0	
Class Data			
Males	45	48	
Females	215	0	

Conclusions: Red/White Eye Color in Drosophila

- Calculate the phenotypic ratios based on the data for males and females separately, and record them in Table 9.5. Do the results differ from the expected ratio per individual data? _____ per class data? _____ If so, explain. _____
_____
_____

- Using the Punnett square provided, calculate the expected phenotypic results for the cross $X^R Y \times X^r X^r$. What is the expected phenotypic ratio among the offspring? males _____ females _____

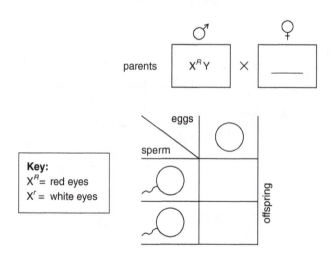

X-Linked Genetics Problems

1. State the genotypes and gametes for each of these fruit flies:

	genotype	gamete(s)
white-eyed male	_____	_____
white-eyed female	_____	_____
red-eyed male	_____	_____
homozygous red-eyed female	_____	_____
heterozygous red-eyed female	_____	_____

2. What are the phenotypic ratios if a white-eyed female is crossed with a red-eyed male?

   males _____ females _____

3. Regardless of any type cross, do white-eyed males inherit the allele for white eyes from their father or

   their mother? _____ Explain your answer. _____

4. In sheep, horns are sex linked; H = horns and h = no horns. Using symbols, what cross do you

   recommend if a farmer wants to produce hornless males? _____ × _____

5. In *Drosophila*, bar eye is sex linked; B = bar eye and b = no bar eye. What are the phenotypic ratios for these crosses?

   bar-eyed male × non–bar-eyed female _____ , _____

   bar-eyed male × heterozygous female _____ , _____

   non–bar-eyed male × heterozygous female _____ , _____

6. A female fruit fly has white eyes. What is the genotype of the father? _____ What could be the

   genotype of the mother? _____ or _____

7. In a cross between fruit flies, all the males have white eyes and the females are 1:1. What is the

   genotype of the parents? female parent _____ male parent _____

8. In a cross between fruit flies, a white-eyed male and a red-eyed female produce no offspring that have

   white eyes. What is the genotype of the parents? male parent _____ female parent _____

9. Make up a sex-linked genetic cross using words and parental genotypes. Create a key and show a
   Punnett square and the phenotypic ratios.

1. If offspring exhibit a 3:1 phenotypic ratio, what are the genotypes of the parents? _____

2. In fruit flies, which of the characteristics you studied was X-linked? _____

3. If offspring exhibit a 9:3:3:1 phenotypic ratio, what are the genotypes of the parental *(P)* generation?

   _____

4. If a cross results in 90 long-winged flies to 30 vestigial-winged flies, what are the phenotypes of the parents?

   _____

5. Briefly describe the life cycle of *Drosophila.* _____

   _____

6. In the cross *AaBb* × *aaBb*, what are the gametes for *AaBb*? _____ for *aaBb*? _____ What are

   the genotypic (not phenotypic) results for this cross? _____

   _____

7. What is the genotype of a white-eyed male fruit fly? _____

8. Suppose you count 40 green tobacco seedlings and 2 white tobacco seedlings in one agar plate. Do your

   results show that both parent plants were heterozygous for the color allele? _____

   Explain your answer. _____

9. Suppose you count tobacco seedlings in six agar plates, and your data are as follows: 125 green plants and

   39 white plants. What is the phenotypic ratio? _____

   _____

10. Suppose that students in the laboratory periods before you removed some of the purple and yellow corn

    kernels from the ears of corn as they were performing the Experimental Procedure. What specific effects

    would this have on your results? _____

    _____

10

DNA Biology and Technology

Learning Outcomes

10.1 DNA Structure and Replication
- Explain how the structure of DNA facilitates replication.
- Explain how DNA replication is semiconservative.

10.2 RNA Structure
- Compare DNA and RNA, discussing their similarities as well as what distinguishes one from the other.

10.3 DNA and Protein Synthesis
- Describe how DNA is able to store the information that specifies a protein.
- Compare the events of transcription with those of translation during protein synthesis.

10.4 Isolation of DNA and Biotechnology
- Describe the procedure for isolating DNA in a test tube.
- Describe the process of DNA gel electrophoresis.

10.5 Detecting Genetic Disorders
- Understand the relationship between an abnormal DNA base sequence and a genetic disorder.
- Suggest two ways to detect a genetic disorder.

Introduction

This laboratory pertains to molecular genetics and biotechnology. Molecular genetics is the study of the structure and function of **DNA (deoxyribonucleic acid),** the genetic material. **Biotechnology** is the manipulation of DNA for the benefit of human beings and other organisms.

First we will study the structure of DNA and see how that structure facilitates DNA replication in the nucleus of cells. DNA replicates prior to cell division; following cell division, each daughter cell has a complete copy of the genetic material. DNA replication is also needed to pass genetic material from one generation to the next. You may have an opportunity to use models to see how replication occurs.

Then we will study the structure of **RNA (ribonucleic acid)** and how it differs from that of DNA before examining how DNA, with the help of RNA, specifies protein synthesis. The linear construction of DNA, in which nucleotide follows nucleotide, is paralleled by the linear construction of the primary structure of protein, in which amino acid follows amino acid. Essentially, we will see that the sequence of nucleotides in DNA codes for the sequence of amino acids in a protein. We will also review the role of three types of RNA in protein synthesis. DNA's code is passed to messenger RNA (mRNA), which moves to the ribosomes containing ribosomal RNA (rRNA). Transfer RNA (tRNA) brings the amino acids to the ribosomes, and they become sequenced in the order directed by mRNA.

We now understand that a mutated gene has an altered DNA base sequence, which can lead to a genetic disorder. You will have an opportunity to carry out a laboratory procedure that detects whether an individual is normal, has sickle-cell disease, or is a carrier.

10.1 DNA Structure and Replication

The structure of DNA lends itself to **replication,** the process that makes a copy of a DNA molecule. DNA replication is a necessary part of chromosome duplication, which precedes cell division. It also makes possible the passage of DNA from one generation to the next.

DNA Structure

DNA is a polymer of nucleotides (Fig. 10.1). Each nucleotide is composed of three molecules: deoxyribose (a 5-carbon sugar), a phosphate, and a nitrogen-containing base.

Figure 10.1 Overview of DNA structure.
a. One nucleotide in which the carbons if ribose are numbered. **b.** Diagram of DNA double helix shows that the molecule resembles a twisted ladder. Sugar-phosphate backbones make up the sides of the ladder, and hydrogen-bonded bases make up the rungs of the ladder. Complementary base pairing dictates that A is bonded to T, G is bonded to C, and vice versa.

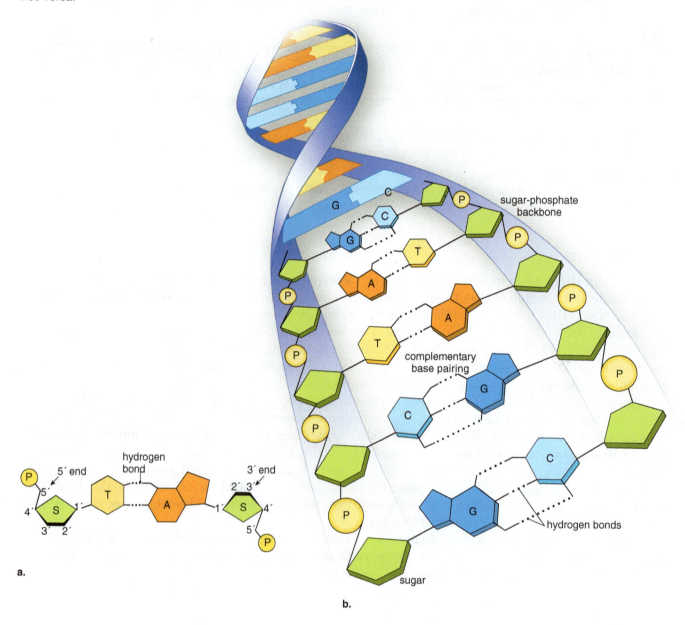

1. A single nucleotide pair is shown in Figure 10.1. If you are working with a kit, draw a representation of one of your nucleotides here. *Label phosphate, base pair, and deoxyribose in your drawing.*

2. Notice the four types of bases: cytosine (C), thymine (T), adenine (A), and guanine (G). What is the color of each of the four types of bases in Figure 10.1? In your kit? Complete Table 10.1 by writing in the colors of the bases.

Table 10.1 Base Colors		
	In Figure 10.1	**In Your Kit**
Cytosine		
Thymine		
Adenine		
Guanine		

3. Using Figure 10.1 as a guide, join several nucleotides together. Observe the entire DNA molecule. What types of molecules make up the backbone (uprights of ladder) of DNA (Fig. 10.1)? _____ and _____. In the backbone, the phosphate of one nucleotide is bonded to a sugar of the next nucleotide.

4. Using Figure 10.1 as a guide, join the bases together with hydrogen bonds. *Label a hydrogen bond in Figure 10.1.* Dashes are used to represent hydrogen bonds in Figure 10.1 because hydrogen bonds are (strong or weak) _____.

5. Notice in Figure 10.1 and in your model that the base A is always paired with the base _____, and the base C is always paired with the base _____. This is called **complementary base pairing.**

6. In Figure 10.1, what molecules make up the rungs of the ladder? _____

7. Each half of the DNA molecule is a DNA strand. Why is DNA also called a double helix (Fig. 10.1)?

_____

DNA Replication

During replication, the DNA molecule is duplicated so that there are two identical DNA molecules. We will see that complementary base pairing makes replication possible.

1. Before replication begins, DNA is unzipped. Using Figure 10.2*a* as a guide, break apart your two DNA strands. What bonds are broken in order to unzip the DNA

   strands? _____

2. Using Figure 10.2*b* as a guide, attach new complementary nucleotides to each strand using complementary base pairing.

3. Show that you understand complementary base pairing by completing Table 10.2.

4. You now have two DNA molecules (Fig. 10.2*c*). Are your molecules identical?

   _____

5. Because of complementary base pairing, each new double helix is composed of an _____ strand

   and a _____ strand. *Write "old"*

   *or "new" in 1 to 10, Figure 10.2*a, b, *and* c. *Conservative* means to save something from the past. Why is DNA replication called semiconservative?

   _____

   _____

   _____

Figure 10.2   DNA replication.
Use of the ladder configuration better illustrates how replication takes place. **a.** The parental DNA molecule. **b.** The "old" strands of the parental DNA molecule have separated. New complementary nucleotides available in the cell are pairing with those of each old strand. **c.** Replication is complete.

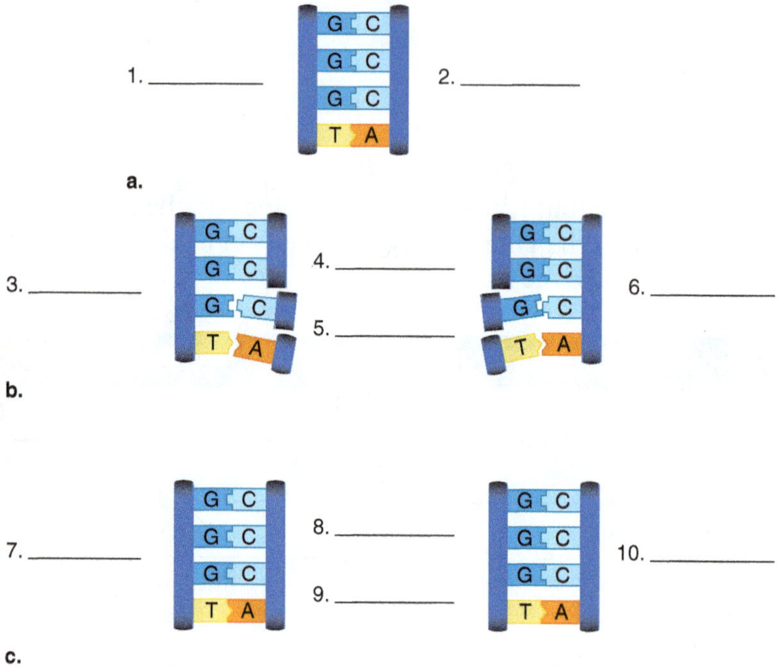

1. _____   2. _____

a.

3. _____   4. _____   6. _____

5. _____

b.

7. _____   8. _____   10. _____

9. _____

c.

6. Genetic material has to be inherited from cell to cell and organism to organism. Consider that, because of DNA replication, a chromosome is composed of two chromatids, and each chromatid is a DNA double helix. The chromatids separate during cell division so that each daughter cell receives a copy of each chromosome. Does replication provide a means for passing DNA from cell to cell and

   organism to organism? _____ Explain. _____

   _____

Table 10.2 DNA Replication																									
Old strand	G	G	G	T	T	C	C	A	T	T	A	A	A	T	T	C	C	A	G	A	A	A	T	C	A	T	A
New strand																											

10.2   RNA Structure

Like DNA, RNA is a polymer of nucleotides (Fig. 10.3). In an RNA nucleotide, the sugar ribose is attached to a phosphate molecule and to a nitrogen-containing base, C, U, A, or G. In RNA, the base uracil replaces thymine as one of the bases. RNA is single stranded, whereas DNA is double stranded.

Figure 10.3   Overview of RNA structure.
RNA is a single strand of nucleotides. *Label the boxed nucleotide as directed in the next Observation.*

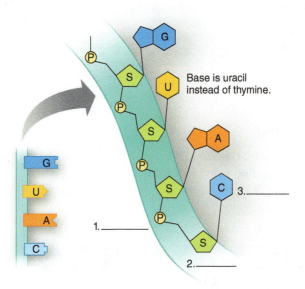

1. Describe the backbone of an RNA molecule. _____
2. Where are the bases located in an RNA molecule? _____

Observation: RNA Structure

1. If you are using a kit, draw a nucleotide from your kit for the construction of mRNA. *Label the ribose (the sugar in RNA), the phosphate, and the base in your drawing and in 1–3, Figure 10.3.*

2. Complete Table 10.3 by writing in the colors of the bases in Figure 10.3 and in your kit.

Table 10.3 Base Colors		
	In Figure 10.3	**In Your Kit**
Cytosine		
Uracil		
Adenine		
Guanine		

3. The base uracil substitutes for the base thymine in RNA. Complete Table 10.4 to show the several other ways RNA differs from DNA.

Table 10.4 DNA Structure Compared with RNA Structure		
	DNA	**RNA**
Sugar	Deoxyribose	
Bases	Adenine, guanine, thymine, cytosine	
Strands	Double stranded with base pairing	
Helix	Yes	

Complementary Base Pairing

Complementary base pairing occurs between DNA and RNA. The RNA base uracil pairs with the DNA base adenine; the other bases pair as studied previously.

Complete Table 10.5 to show the complementary DNA bases for the RNA bases.

Table 10.5 Complementary Base Pairing Between DNA and RNA				
RNA bases	C	U	A	G
DNA bases				

10.3 DNA and Protein Synthesis

Protein synthesis requires the processes of transcription and translation. During **transcription,** which takes place in the nucleus, an RNA molecule called **messenger RNA (mRNA)** is made complementary to one of the DNA strands. This mRNA leaves the nucleus and goes to the ribosomes in the cytoplasm. Ribosomes are composed of **ribosomal RNA (rRNA)** and proteins in two subunits.

During **translation,** RNA molecules called **transfer RNA (tRNA)** bring amino acids to the ribosome, and they join in the order prescribed by mRNA. This sequence of amino acids was originally specified by DNA. This is the information that DNA, the genetic material, stores.

What is the role of each of these participants in protein synthesis?

DNA _____

mRNA _____

tRNA _____

Transcription

During transcription, complementary RNA is made from a DNA template (Fig. 10.4). A portion of DNA unwinds and unzips at the point of attachment of the enzyme **RNA polymerase.** A strand of mRNA is produced when complementary nucleotides join in the order dictated by the sequence of bases in DNA. Transcription occurs in the nucleus, and the mRNA passes out of the nucleus to enter the cytoplasm.

Label Figure 10.4. For number 1, note the name of the enzyme that carries out mRNA synthesis. For number 2, note the name of this molecule.

Observation: Transcription

1. If you are using a kit, unzip your DNA model so that only one strand remains. This strand is the **sense strand,** the strand that is transcribed.
2. Using Figure 10.4 as a guide, construct a messenger RNA (mRNA) molecule by first lining up RNA nucleotides complementary to the sense strand of your DNA molecule. Join the RNA nucleotides together to form mRNA.
3. A portion of DNA has the sequence of bases shown in Table 10.6. *Complete Table 10.6 to show the sequence of bases in mRNA.*
4. If you are using a kit, unzip mRNA transcript from the DNA. Locate the end of the strand that will move to

    the _____ in the cytoplasm.

Figure 10.4   Messenger RNA (mRNA).
Messenger RNA complementary to a section of DNA forms during transcription.

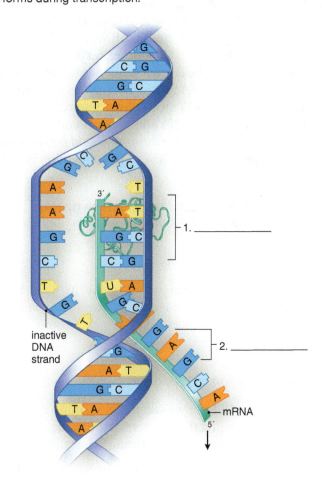

1. _____

2. _____

inactive DNA strand

mRNA

Table 10.6	Transcription
DNA	T A C A C G A G C A A C T A A C A T
mRNA	

Translation

DNA specifies the sequence of amino acids in a polypeptide because every three bases code for an amino acid. Therefore, DNA is said to have a **triplet code.** The bases in mRNA are complementary to the bases in DNA. Every three bases in mRNA are called a **codon.** One codon of mRNA represents one amino acid. Thus, the sequence of DNA bases serves as the blueprint for the sequence of amino acids assembled to make a protein. The correct sequence of amino acids in a polypeptide is the message that mRNA carries.

Messenger RNA leaves the nucleus and proceeds to the ribosomes, where protein synthesis occurs. Transfer RNA (tRNA) molecules are so named because they transfer amino acids to the ribosomes. Each RNA has a specific tRNA amino acid at one end and a matching **anticodon** at the other end (Fig. 10.5). *Label Figure 10.5, where the amino acid is represented as a colored ball, the tRNA is green, and the anticodon is the sequence of three bases. (The anticodon is complementary to the mRNA codon.)*

Figure 10.5   Transfer RNA (tRNA).
Transfer RNA carries amino acids to the ribosomes.

1. _____
2. _____
3. _____

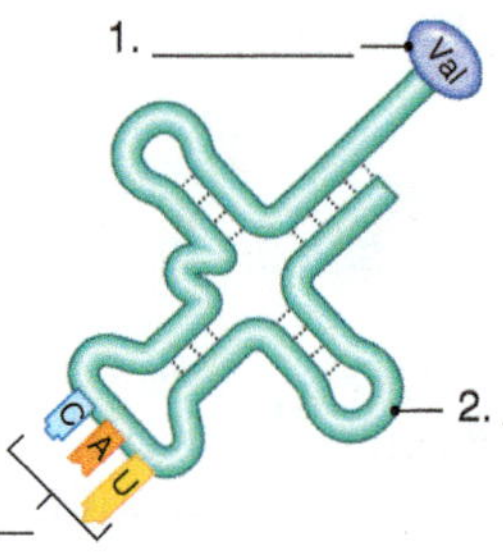

Observation: Translation

1. Figure 10.6 shows seven tRNA–amino acid complexes. Every amino acid has a name; in the figure, only the first three letters of the name are inside the ball.
2. If you are using a kit, arrange your tRNA–amino acid complexes in the order consistent with

   Table 10.7. *Complete Table 10.7.* Why are the codons and anticodons in groups of three? _____
   _____

Figure 10.6   Transfer RNA diversity.
Each type of tRNA carries only one particular amino acid, designated here by the first three letters of its name.

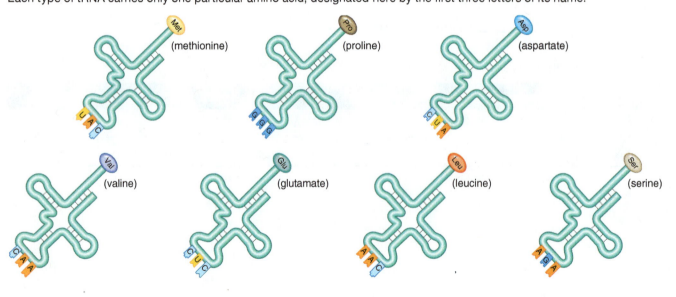

Table 10.7 Translation							
mRNA codons	AUG	CCC	GAG	GUU	GAU	UUG	UCU
tRNA anticodons							
Amino acid*							

*Use three letters only. See Table 10.8 for the full names of these amino acids.

Table 10.8 Names of Amino Acids	
Abbreviation	**Name**
Met	Methionine
Pro	Proline
Asp	Aspartate
Val	Valine
Glu	Glutamate
Leu	Leucine
Ser	Serine

3. Figure 10.7 shows the manner in which the polypeptide grows. A ribosome has room for two tRNA complexes at a time. As the first tRNA leaves, it passes its amino acid or peptide to the next tRNA–amino acid complex. Then the ribosome moves forward, making room for the next tRNA–amino acid. This sequence of events occurs over and over until the entire polypeptide is borne by the last tRNA to come to the ribosome. Then a release factor releases the polypeptide chain from the ribosome. *In Figure 10.7, label the ribosome, and the mRNA.*

Figure 10.7 Protein synthesis.
1. A ribosome has a binding site for two tRNA–amino acid complexes. **2.** Before a tRNA leaves, an RNA passes its attached peptide to a newly arrived tRNA–amino acid complex. **3.** The ribosome moves forward, and the next tRNA–amino acid complex arrives.

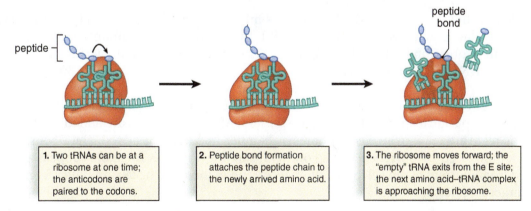

1. Two tRNAs can be at a ribosome at one time; the anticodons are paired to the codons.

2. Peptide bond formation attaches the peptide chain to the newly arrived amino acid.

3. The ribosome moves forward; the "empty" tRNA exits from the E site; the next amino acid–tRNA complex is approaching the ribosome.

10.4 Isolation of DNA and Biotechnology

In the following experiment, you will isolate DNA from the cells of a fruit or vegetable. It will only be necessary to expose the cells to an agent (dishwasher detergent) that emulsifies membrane in order to "free" the DNA from its enclosures (plasma membrane and nuclear envelope). When transferred to a tube, the presence of NaCl allows DNA to precipitate as a sodium salt. The precipitate forms at the interface between ethanol and the salt solution, and then it floats to the top of the tube, where it may be collected.

Experimental Procedure: Isolation of DNA

1. Acquire a slice of fruit or vegetable (i.e., tomato, onion, or apple), a mortar and pestle, and a large, clean glass test tube on ice.

2. Crush and grind the slice of fruit or vegetable in a mortar and pestle. Remove the pestle and set aside. Add enough 0.9% NaCl solution to achieve a "soupy" consistency.

3. Add two drops of dishwasher detergent (such as Blue Dawn) to the mixture in the mortar. Swirl until the color of the detergent disappears. Wait 3 to 5 minutes. The solution becomes clear and viscous as the DNA escapes its enclosures.

4. Check the viscosity of the solution in the mortar, and if it is too viscous for you to pipet, add a few more ml of NaCl solution. Use a transfer pipet to move 2 to 3 ml of the DNA from the mortar to the glass tube. Try to pick up clear zones of the solution WITHOUT CELL DEBRIS.

5. Slowly add 5 to 7 ml of ice-cold ethanol down the side of the tube on ice. As you do so, DNA will precipitate *between* the ethanol and the NaCl solutions. Then it will float to the top, where it can be picked up by a fresh transfer pipet. Any "white dust" you see consists of RNA molecules.

6. Use a transfer pipet to place the DNA in a small, clean test tube. Pipet away any extra water/ethanol and air-dry the DNA for a few minutes. Dissolve the DNA in 3 to 4 ml distilled H_2O.

7. Add five drops of phenol red, a pH indicator. The resulting dark pink color confirms the presence of nucleic acid (i.e., the DNA).

> ⚠ For experimental procedures, wear safety goggles, gloves, and protective clothing. If any chemical spills on your skin, wash immediately with mild soap and water; flood eyes with water only. Report any spills immediately to your instructor.

Experimental Procedure: Gel Electrophoresis

During gel electrophoresis, charged DNA molecules migrate across a span of gel (gelatinous slab) because they are placed in a powerful electrical field. In the present experiment, each DNA sample is placed in a small depression in the gel called a well. The gel is placed in a powerful electrical field. The electricity causes DNA fragments, which are negatively charged, to move through the gel according to their size.

Almost all DNA gel electrophoresis is carried out using horizontal gel slabs (Fig. 10.8). First, the gel is poured onto a plastic plate, and the wells are formed. After the samples are added to the wells, the gel and the plastic plate are put into an electrophoresis chamber, and buffer is added. The DNA samples begin to migrate after the electrical current is turned on. With staining, the DNA fragments appear as a series of bands spread from one end of the gel to the other according to their size because smaller fragments move faster than larger fragments.

Figure 10.8 Equipment and procedure for gel electrophoresis.

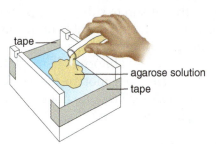

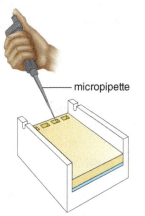

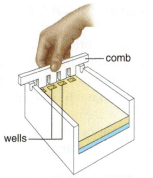

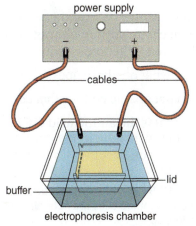

a. Agarose solution poured into casting tray

b. Comb that forms wells for samples

c. Wells that can be loaded with samples

d. Electrophoresis chamber and power supply

Answer the following questions:

1. What is biotechnology? (See page 97.) _____

_____

_____

2. Speculate how the ability to isolate DNA and run gel electrophoresis of DNA relates to biotechnology.

_____

_____

_____

3. Name a biotechnology product someone you know is now using or taking as a medicine.

_____

10.5 Detecting Genetic Disorders

The base sequence of DNA in all the chromosomes is an organism's genome. Now that the Human Genome Project is finished, we know the usual order of all the 3.2 billion nucleotide base pairs in the haploid human genome. Today it is possible to sequence anyone's genome within a relatively short time, and thereby determine what particular base sequence alterations signify that he or she has a disorder or will have one in the future. In this laboratory, you will study the alteration in base sequence that causes a person to have sickle-cell disease.

In persons with sickle-cell disease, the red blood cells aren't biconcave disks like normal red blood cells—they are sickle-shaped. Sickle-shaped cells can't pass along narrow capillary passageways. They clog the vessels and break down, causing the person to suffer from poor circulation, anemia, and poor resistance to infection. Internal hemorrhaging leads to further complications, such as jaundice, episodic pain in the abdomen and joints, and damage to internal organs.

Sickle-shaped red blood cells are caused by an abnormal hemoglobin (*Hb^S*). Individuals with the *Hb^A Hb^A* genotype are normal; those with the *Hb^S Hb^S* genotype have sickle-cell disease, and those with the *Hb^A Hb^S* have sickle-cell trait. Persons with sickle-cell trait do not usually have sickle-shaped cells unless they experience dehydration or mild oxygen deprivation.

Genetic Sequence for Sickle-Cell Disease

Examine Figure 10.9*a* and *b,* which shows the DNA base sequence, the mRNA codons, and the amino acid sequence for a portion of the gene for *Hb^A* and the same portion for *Hb^S*. Today many genetic disorders can be detected by genomic sequencing.

1. In what one base does *Hb^A* differ from *Hb^S*?   *Hb^A* _____   *Hb^S* _____

2. What are the codons that contain this base?   *Hb^A* _____   *Hb^S* _____

3. What is the amino acid difference?   *Hb^A* _____   *Hb^S* _____

Figure 10.9   Sickle-cell disease.
a. When red blood cells are normal, the base sequence (in one location) for *Hb^A* alleles is CTC. **b.** In sickle-cell disease at these locations, it is CAC.

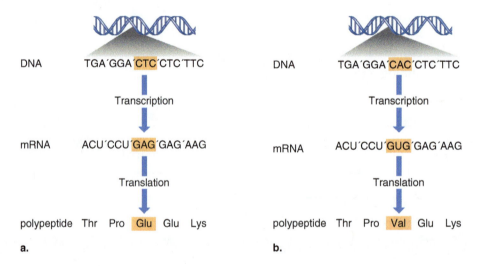

This amino acid difference causes the polypeptide chain in sickle-cell hemoglobin to pile up as firm rods that push against the plasma membrane and deform the red blood cell into a sickle shape:

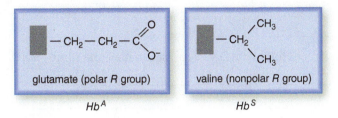

glutamate (polar *R* group)   valine (nonpolar *R* group)

Hb^A Hb^S

Detection of Sickle-Cell Disease by Gel Electrophoresis

Three samples of hemoglobin have been subjected to protein gel electrophoresis. Protein gel electrophoresis is carried out in the same manner as DNA gel electrophoresis (see Fig. 10.10), except the gel has a different composition.

1. Sickle-cell hemoglobin *(Hb^S)* migrates slower toward the positive pole than normal hemoglobin *(Hb^A)* because the amino acid valine has no polar *R* groups, whereas the amino acid glutamate does have a polar *R* group.

2. In Figure 10.10, which lane contains only *Hb^S*, signifying that the individual is *Hb^{S}Hb^S?* _____

3. Which lane contains only *Hb^A*, signifying that the individual is *Hb^{A}Hb^A?* _____

4. Which lane contains both *Hb^S* and *Hb^A*, signifying that the individual is *Hb^{A}Hb^S?* _____

Figure 10.10   Gel electrophoresis of hemoglobins.

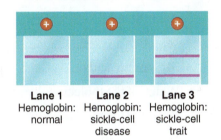

| Lane 1 | Lane 2 | Lane 3 |
| Hemoglobin: normal | Hemoglobin: sickle-cell disease | Hemoglobin: sickle-cell trait |

Detection by Genomic Sequencing

You are a genetic counselor. A young couple seeks your advice because sickle-cell disease occurs among the family members of each. You order DNA base sequencing to be done. The results come back that, at one of the loci for normal hemoglobin, each has the abnormal sequence CAC instead of CTC. The other locus is normal. What are the chances that this couple will have a child with sickle-cell disease? _____

Summary: Detecting Genetic Disorders

• What two methods of detecting sickle-cell disease were described in this section? _____
_____

• Which method is more direct and probably requires more expensive equipment to do? _____
_____

• Which method probably preceded the other method as a means to detect sickle-cell disease? _____
_____

Laboratory Review 10

1. Explain why DNA is said to have a structure that resembles a ladder. _____
_____

2. Do the two DNA double helices following DNA replication have the same, or a different, composition?
_____
_____

3. How is complementary base pairing different when pairing DNA to DNA than when pairing DNA to mRNA?
_____
_____

4. Explain why the genetic code is called a triplet code. _____

5. What role does each of the following molecules play in protein synthesis?

   a. DNA _____

   b. mRNA _____

   c. tRNA _____

   d. Amino acids _____

6. Which of the molecules listed in question 5 are involved in transcription? _____

7. Which of the molecules listed in question 5 are involved in translation? _____

8. What is the purpose of gel electrophoresis? _____
_____

9. Why does sickle-cell hemoglobin *(Hb^S)* migrate slower than normal hemoglobin *(Hb^A)* during gel

   electrophoresis? _____

10. Below is a sequence of bases associated with the template DNA strand:
    TAC CCC GAG CTT

    a. Identify the sequence of bases in the mRNA resulting from the transcription of the above DNA
       sequence.

    b. Identify the sequence of bases in the tRNA anticodon that will bind with the first codon on the mRNA
       identified above.

Essentials of Biology Website

Instructors can find lab prep information and answers to all of the laboratory questions in the Laboratory Resource Guide. *Students* can practice their knowledge with quizzes, animations, flashcards, and much more.

www.mhhe.com/maderessentials4

McGraw-Hill Access Science Website

An online encyclopedia of science and technology that provides information, including videos, that can enhance the laboratory experience.

www.accessscience.com

LEARNSMART

DNA & Biotechnology

11

Genetic Counseling

Introduction

In this laboratory, you will discover that the same principles of genetics apply to humans as they do to plants and fruit flies. A gene has two alternate forms, called **alleles,** for any trait, such as hairline, finger length, and so on. One possible allele, designated by a capital letter, is **dominant** over the **recessive** allele, designated by a lowercase letter. An individual can be **homozygous dominant** (two dominant alleles, *EE*), **homozygous recessive** (two recessive alleles, *ee*), or **heterozygous** (one dominant and one recessive allele, *Ee*). **Genotype** refers to an individual's alleles, and **phenotype** refers to an individual's appearance (Fig. 11.1). Homozygous dominant and heterozygous individuals show the dominant phenotype; homozygous recessive individuals show the recessive phenotype.

Figure 11.1 Genotype versus phenotype.
Unattached earlobes *(E)* are dominant over attached earlobes *(e)*. **a.** Homozygous dominant individuals have unattached earlobes.
b. Homozygous recessive individuals have attached earlobes. **c.** Heterozygous individuals have unattached earlobes.

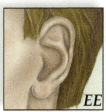
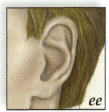
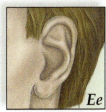

a. Unattached earlobe   b. Attached earlobe   c. Unattached earlobe

11.1 Determining the Genotype

Humans inherit 46 chromosomes, which occur in 23 pairs. Twenty-two of these pairs are called autosomes and 1 pair is the sex chromosomes. Autosomal traits are determined by alleles on the autosomal chromosomes.

Autosomal Dominant and Recessive Traits

Using the alleles from Table 11.1, answer these questions.

1. What is the homozygous dominant genotype for type of hairline? _____

   What is the phenotype? _____

2. What is the homozygous recessive genotype for finger length? _____

   What is the phenotype? _____

3. Why does the heterozygous individual *Ff* have freckles? _____

These genetic problems also use the alleles from Table 11.1.

4. Maria and the members of her immediate family have attached earlobes. What is Maria's

   genotype? _____

   Her maternal grandfather has unattached earlobes. Deduce the genotype of her maternal

   grandfather. _____

   Explain your answer. _____

   _____

5. Moses does not have a bent little finger, but his parents do. Deduce the genotype of his

   parents. _____

   Deduce the genotype of Moses. _____

   Explain your answer. _____

   _____

6. Manny is adopted. He has hair on the back of his hand. Could both of his parents have had hair on the

   back of the hand? _____

   Could both of his parents have had no hair on the back of the hand? _____

   Explain your answer. _____

   _____

7. Simona and her husband have widow peaks. One child has a widow's peak and the other does not.

   Give the possible genotypes of all persons involved.

   Simona and her husband _____

   Child with straight hairline _____

   Child with widow's peak _____ or _____

Experimental Procedure: Human Traits

1. For this Experimental Procedure, you will need a lab partner to help you determine your phenotype for the traits listed in the first column of Table 11.1.

2. Determine your probable genotype. If you have the recessive phenotype, you know your genotype. If you have the dominant phenotype, you may be able to decide whether you are homozygous dominant or heterozygous by recalling the phenotype of your parents, siblings, or children. Circle your probable genotype in the second column of Table 11.1.

3. Your instructor will tally the class's phenotypes for each trait so that you can complete the third column of Table 11.1.

4. Complete Table 11.1 by calculating the percentage of the class with each trait. Are dominant phenotypes always the most common in a population? _____

   Explain your answer. _____

   _____

Table 11.1 Autosomal Human Traits

Trait: d = Dominant r = Recessive	Probable Genotypes	Number in Class	Percentage of Class with Trait
Hairline:			
Widow's peak (d)	*WW* or *Ww*	_____	_____
Straight hairline (r)	*ww*	_____	_____
Earlobes:			
Unattached (d)	*UU* or *Uu*	_____	_____
Attached (r)	*uu*	_____	_____
Skin pigmentation:			
Freckles (d)	*FF* or *Ff*	_____	_____
No freckles (r)	*ff*	_____	_____
Hair on back of hand:			
Present (d)	*HH* or *Hh*	_____	_____
Absent (r)	*hh*	_____	_____
Thumb hyperextension—"hitchhiker's thumb":			
Last segment cannot be bent backward (d)	*TT* or *Tt*	_____	_____
Last segment can be bent back to 60° (r)	*tt*	_____	_____
Bent little finger:			
Little finger bends toward ring finger (d)	*LL* or *Ll*	_____	_____
Straight little finger (r)	*ll*	_____	_____
Interlacing of fingers:			
Left thumb over right (d)	*II* or *Ii*	_____	_____
Right thumb over left (r)	*ii*	_____	_____

11.2 Patterns of Genetic Inheritance

Recall that a Punnett square is a means to determine the genetic inheritance of offspring if the genotypes of both parents are known. In a **Punnett square**, all possible types of sperm are lined up vertically, and all possible types of eggs are lined up horizontally, or vice versa, so that every possible combination of gametes occurs within the square. Figure 11.2 shows how to construct a Punnett square when autosomal alleles are involved.

Figure 11.2   Punnett square.
In a Punnett square, all possible sperm are displayed vertically and all possible eggs are displayed horizontally, or vice versa. The genotypes of the offspring (in this case, also the phenotypes) are in the squares.

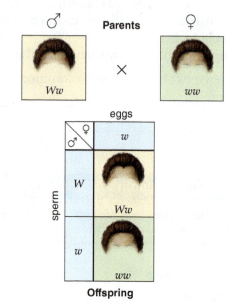

Results of Cross
Phenotypic ratio 1:1
☐ Chance of widow's peak ½ = 50%
☐ Chance of straight hairline ½ = 50%

Inheritance of Genetic Disorders

Figure 11.3 can be used to learn the chances of a particular phenotype.

In Figure 11.3*a*,

¼ of the offspring have the recessive phenotype = _____ % chance

¾ of the offspring have the dominant phenotype = _____ % chance

In Figure 11.3*b*,

½ of the offspring have the recessive or the dominant phenotype = _____ % chance

In all the following genetic problems, use letters to fill in the parentheses with the genotype of the parents.

1. **a.** With reference to Figure 11.3*a*, if a genetic disorder is recessive and both parents are heterozygous (_____), what are the chances that an offspring will have the disorder? _____
    b. With reference to Figure 11.3*a*, if a genetic disorder is dominant and both parents are heterozygous (_____), what are the chances that an offspring will have the disorder? _____

2. **a.** With reference to Figure 11.3*b*, if the parents are heterozygous (_____) by homozygous recessive (_____), and the genetic disorder is recessive, what are the chances that the offspring will have the disorder? _____
    b. With reference to Figure 11.3*b*, if the parents are heterozygous (_____) by homozygous recessive (_____), and the genetic disorder is dominant, what are the chances that an offspring will have the disorder? _____

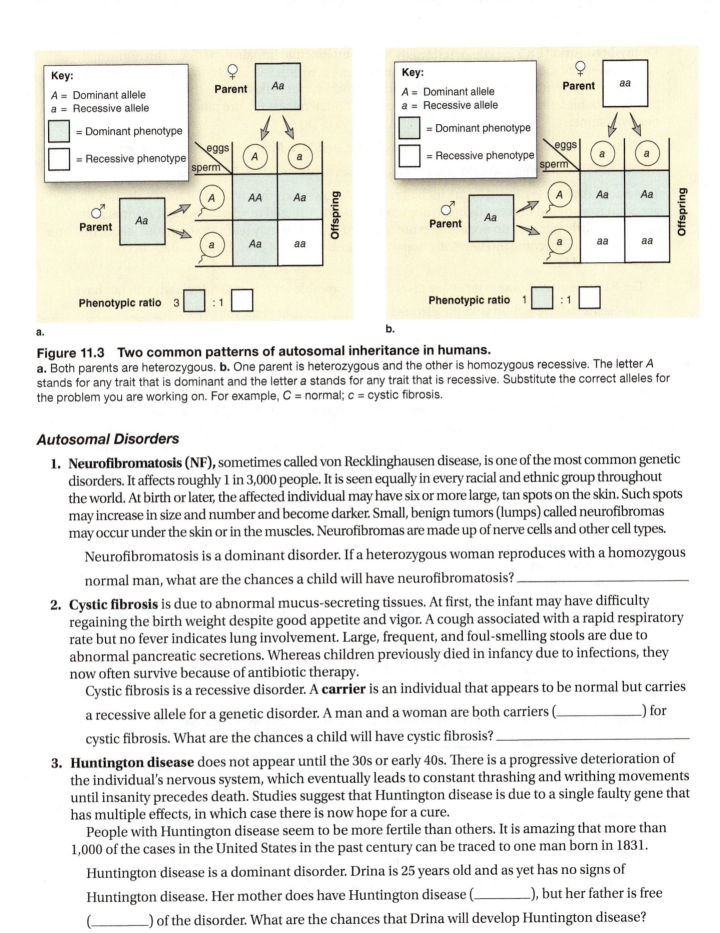

Figure 11.3 Two common patterns of autosomal inheritance in humans.
a. Both parents are heterozygous. **b.** One parent is heterozygous and the other is homozygous recessive. The letter *A* stands for any trait that is dominant and the letter *a* stands for any trait that is recessive. Substitute the correct alleles for the problem you are working on. For example, *C* = normal; *c* = cystic fibrosis.

Autosomal Disorders

1. **Neurofibromatosis (NF),** sometimes called von Recklinghausen disease, is one of the most common genetic disorders. It affects roughly 1 in 3,000 people. It is seen equally in every racial and ethnic group throughout the world. At birth or later, the affected individual may have six or more large, tan spots on the skin. Such spots may increase in size and number and become darker. Small, benign tumors (lumps) called neurofibromas may occur under the skin or in the muscles. Neurofibromas are made up of nerve cells and other cell types.

   Neurofibromatosis is a dominant disorder. If a heterozygous woman reproduces with a homozygous normal man, what are the chances a child will have neurofibromatosis? _____

2. **Cystic fibrosis** is due to abnormal mucus-secreting tissues. At first, the infant may have difficulty regaining the birth weight despite good appetite and vigor. A cough associated with a rapid respiratory rate but no fever indicates lung involvement. Large, frequent, and foul-smelling stools are due to abnormal pancreatic secretions. Whereas children previously died in infancy due to infections, they now often survive because of antibiotic therapy.

   Cystic fibrosis is a recessive disorder. A **carrier** is an individual that appears to be normal but carries a recessive allele for a genetic disorder. A man and a woman are both carriers (_____) for cystic fibrosis. What are the chances a child will have cystic fibrosis? _____

3. **Huntington disease** does not appear until the 30s or early 40s. There is a progressive deterioration of the individual's nervous system, which eventually leads to constant thrashing and writhing movements until insanity precedes death. Studies suggest that Huntington disease is due to a single faulty gene that has multiple effects, in which case there is now hope for a cure.

   People with Huntington disease seem to be more fertile than others. It is amazing that more than 1,000 of the cases in the United States in the past century can be traced to one man born in 1831.

   Huntington disease is a dominant disorder. Drina is 25 years old and as yet has no signs of Huntington disease. Her mother does have Huntington disease (_____), but her father is free (_____) of the disorder. What are the chances that Drina will develop Huntington disease? _____

4. **Phenylketonuria (PKU)** is characterized by severe intellectual impairment due to an abnormal accumulation of the common amino acid phenylalanine within cells, including neurons. The disorder takes its name from the presence of a breakdown product, phenylketone, in the urine and blood. Newborn babies are routinely tested at the hospital and, if necessary, are placed on a diet low in phenylalanine.

Phenylketonuria (PKU) is a recessive disorder. Mr. and Mrs. Martinez appear to be normal, but they have a child with PKU. What are the genotypes of Mr. and Mrs. Martinez? _____

5. **Tay–Sachs disease** is caused by the inability to break down a certain type of fat molecule that accumulates around nerve cells until they are destroyed. Afflicted newborns appear normal and healthy at birth, but they do not develop normally. At first, they may learn to sit up and stand, but later they regress and become intellectually impaired, blind, and paralyzed. Death usually occurs between ages three and four.

Tay–Sachs is an autosomal recessive disorder. Is it possible for two individuals who do not have Tay–Sachs to have a child with the disorder? _____ Explain your answer. _____

_____

X-Linked Disorders

The sex chromosomes designated X and Y carry genes just like the autosomal chromosomes. Some genes, particularly on the X chromosome, have nothing to do with gender inheritance and are said to be X-linked. **X-linked recessive disorders** are due to recessive genes carried on the X chromosomes. Males are more likely to have an X-linked recessive disorder than females because the Y chromosome is blank for this trait. Does a color-blind male give his son a recessive-bearing X or a Y that is blank for the recessive allele? _____

The possible genotypes and phenotypes for an X-linked recessive disorder are as follows:

Females	Males
$X^B X^B$ = normal vision	$X^B Y$ = normal vision
$X^B X^b$ = normal vision (carrier)	$X^b Y$ = color blindness
$X^b X^b$ = color blindness	

An X-linked recessive disorder in a male is always inherited from his mother. Most likely, his mother is heterozygous and therefore does not show the disorder. She is designated a carrier for the disorder. Figure 11.4 shows how females can become carriers.

1. **a.** What is the genotype for a color-blind female? _____ How many recessive alleles does a female inherit to be color blind? _____

   b. What is the genotype for a color-blind male? _____ How many recessive alleles does a male inherit to be color blind? _____

2. **a.** With reference to Figure 11.4a, if the mother is a carrier (_____) and the father has normal vision (_____), what are the chances that a daughter will be color blind? _____

   b. A daughter will be a carrier? _____ **c.** A son will be color blind? _____

3. **a.** With reference to Figure 11.4b, if the mother has normal vision (_____) and the father is color blind (_____), what are the chances that a daughter will be color blind? _____

   b. A daughter will be a carrier? _____ **c.** A son will be color blind? _____

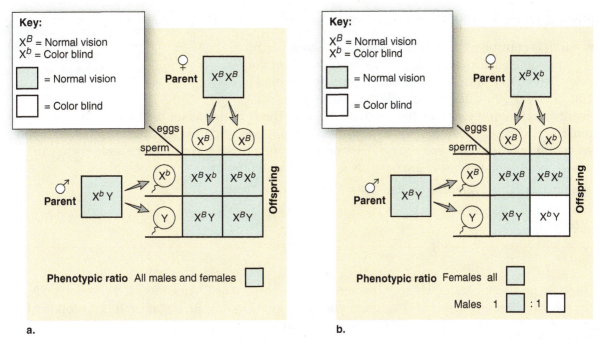

Figure 11.4 Two common patterns of X-linked inheritance in humans.
a. The sons of a carrier mother have a 50% chance of being color blind. **b.** A color-blind father has carrier daughters.

X-Linked Genetics Problems

For **color blindness,** there are two possible X-linked alleles involved. One affects the green-sensitive cones, whereas the other affects the red-sensitive cones. About 6% of men in the United States are color blind due to a mutation involving green perception, and about 2% are color blind due to a mutation involving red perception.

1. A woman with normal color vision (_____), whose father was color blind (_____),

   marries a man with normal color vision (_____). What genotypes could occur among their

   offspring? _____

   What genotypes could occur if it was the normal-visioned man's father who was color blind?

   _____

2. Antonio's father is color blind (_____) but his mother is not color blind (_____).

   Is Antonio necessarily color blind? _____ Explain. _____

   Could he be color blind? _____ Explain. _____

   Hemophilia is called the bleeder's disease because the affected person's blood is unable to clot. Although hemophiliacs do bleed externally after an injury, they also suffer from internal bleeding, particularly around joints. Hemorrhages can be checked with transfusions of fresh blood (or plasma) or concentrates of the clotting protein. The most common type of hemophilia is hemophilia A, due to absence or minimal presence of a particular clotting factor called factor VIII.

3. Make up a cross involving hemophilia that could be answered by a Punnett square, as in

   Figure 11.4*a* or *b*. _____

   _____

   What is the answer to your genetics problem? _____

Multiple Alleles

When a trait is controlled by **multiple alleles,** the gene has more than two possible alleles. But each person has only two of the possible alleles. For example, ABO blood type is determined by multiple alleles: I^A, I^B, i. Red blood cells have surface molecules called antigens that indicate they belong to the person. The I^A allele causes red blood cells to carry an A antigen, the I^B allele causes red blood cells to carry a B antigen, and the i allele causes the red blood cells to have neither of these antigens. I^A and I^B are dominant to i. Remembering that each person can have any two of the possible alleles, these are possible genotypes and phenotypes for blood types.

Genotypes	Antigens on Red Cells	Blood Types
$I^A I^A$, $I^A i$	A	A
$I^B I^B$, $I^B i$	B	B
$I^A I^B$	A and B	AB
ii	none	O

Blood type also indicates whether the person is Rh positive or Rh negative. If the genotype is DD or Dd, the person is Rh positive, and if the genotype is dd, the person is Rh negative. It is customary to simply attach a $+$ or $-$ superscript to the ABO blood type, as in A⁻.

> ⚠️ **Protective clothing** Wear protective laboratory clothing, latex gloves, and goggles. If the chemicals touch the skin, eyes, or mouth, wash immediately. If inhaled, seek fresh air.

Experimental Procedure: Using Blood Type to Help Determine Paternity

In this Experimental Procedure a mother, Wanda, is seeking support for her child, Sophia. We will use blood typing to decide which of three men could be the father.

1. Obtain three testing plates, each of which contains three depressions; vials of blood from possible fathers 1, 2, and 3, respectively; vials of anti-A serum, anti-B serum, and anti-Rh serum. (All of these are synthetic.)
2. Using a wax pencil, number the plates so you know which plate is for possible father 1, 2, or 3. Look carefully at a plate and notice the wells are designated as A, B, or Rh.
3. Being sure to close the cap to each vial in turn, do the following using plate #1:
   Add a drop of father 1 blood to all three wells—close the cap.
   Add a drop of anti-A (blue) to the well designated A—close the cap.
   Add a drop of anti-B (yellow) to the well designated B—close the cap.
   Add a drop of anti-Rh (clear) to the well designated Rh—close the cap.
4. Stir the contents of each well with a mixing stick of the correct color. After a few minutes, examine the wells for agglutination (i.e., granular appearances that indicate the blood type). (Rh⁺ takes the longest to react.) If a person had AB⁺ blood, which wells would show agglutination? _____

   _____

5. Repeat steps 3 and 4 for plates 2 and 3.
6. Record the blood type results for each of the men in Table 11.2.

Table 11.2	Blood Types of Involved Persons				
	Mother*	**Child***		**Father?**	
	Wanda	Sophia	#1	#2	#3
Blood type	B⁻	AB⁺			

*Your instructor may have you confirm these results.

Conclusion

1. Noting that only father 3 could have given Sophia the Rh antigen, from whom did she receive the I^B allele? _____ From which parent did she receive the I^A allele? _____ Is there any other possible interpretation to the results of blood typing? _____

Blood Typing Problems

1. A man with type A blood reproduces with a woman who has type B blood. Their child has blood type O. Using I^A, I^B, and i, give the genotype of all persons involved.

   man _____ woman _____ child _____

2. If a child has type AB blood and the father has type B blood, what could the genotype of the mother be?
   _____ or _____

3. If both mother and father have type AB blood, they cannot be the parents of a child who has what blood type? _____

4. What blood types are possible among the children if the parents are $I^A i \times I^B i$? (*Hint:* Do a Punnett square using the possible gametes for each parent.)

   Punnett Square:

11.3 Genetic Counseling

Potential parents are becoming aware that many illnesses are caused by abnormal chromosome inheritance or by gene mutations. Therefore, they are seeking genetic counseling, which is available in many major hospitals. The counselor helps the couple understand the mode of inheritance for a condition of concern so that the couple can make an informed decision about how to proceed.

Determining Chromosome Inheritance

If a genetic counselor suspects that a condition is due to a chromosome anomaly, he or she may suggest that the chromosome inheritance be examined. It is possible to view the chromosomes of an individual because cells can be microscopically examined and photographed just before cell division occurs. A computer is then used to arrange the chromosomes by pairs. The resulting pattern of chromosomes is called a **karyotype.**

A trisomy occurs when the individual has three chromosomes instead of two chromosomes at one karyotype location. **Trisomy 21** (Down syndrome) is the most common autosomal trisomy in humans. Survival to adulthood is common. Characteristic facial features include an eyelid fold, a flat face, and a large, fissured tongue. Some degree of intellectual impairment is common, as is early-onset Alzheimer disease. Sterility due to sexual underdevelopment may be present.

Observation: Sex Chromosome Anomalies

A female with **Turner syndrome** (XO) has only one sex chromosome, an X chromosome; the O signifies the absence of the second sex chromosome. Because the ovaries never become functional, these females do not undergo puberty or menstruation, and their breasts do not develop. Generally, females with Turner syndrome have a short build, folds of skin on the back of the neck, difficulty recognizing various spatial patterns, and normal intelligence. With hormone supplements, they can lead fairly normal lives.

When an egg having two X chromosomes is fertilized by an X-bearing sperm, an individual with **poly-X syndrome** results. The body cells have 3 X chromosomes and therefore 47 chromosomes. Although they tend to have learning disabilities, poly-X females have no apparent physical anomalies, and many are fertile and have children with a normal chromosome count.

When an egg having two X chromosomes is fertilized by a Y-bearing sperm, a male with **Klinefelter syndrome** results. This individual is male in general appearance, but the testes are underdeveloped, and the breasts may be enlarged. The limbs of XXY males tend to be longer than average, muscular development is poor, body hair is sparse, and many XXY males have learning disabilities.

Jacob syndrome occurs in males who are usually taller than average, suffer from persistent acne, and tend to have speech and reading problems. At one time, it was suggested that XYY males were likely to be criminally aggressive, but the incidence of such behavior has been shown to be no greater than that among normal XY males.

Label each karyotype in Figure 11.5 as one of the syndromes just discussed. Explain your answers on the lines provided.

Figure 11.5 Sex chromosome anomalies.

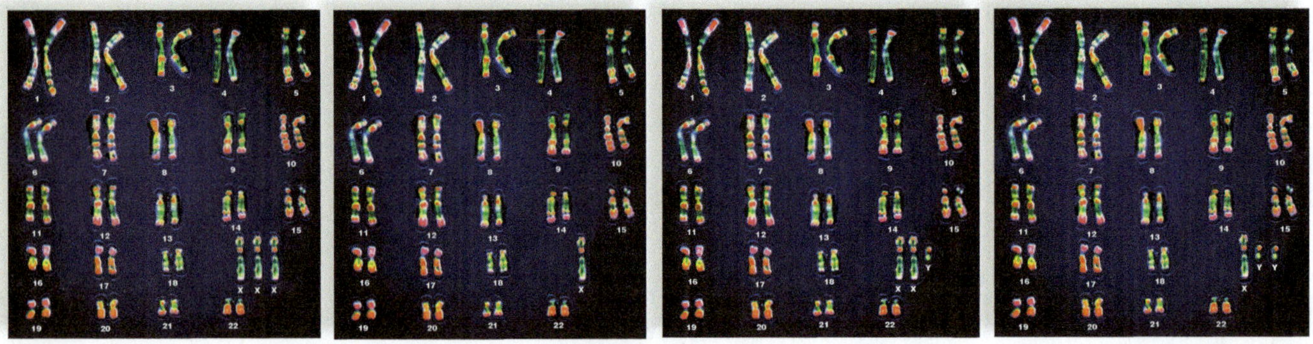

a. _____   b. _____   c. _____   d. _____

Determining the Pedigree

A pedigree shows the inheritance of a genetic disorder within a family and can help determine the inheritance pattern and whether any particular individual has an allele for that disorder. Then a Punnett square can be done to determine the chances of a couple producing an affected child.

The symbols used to indicate normal and affected males and females, reproductive partners, and siblings in a pedigree are shown in Figure 11.6.

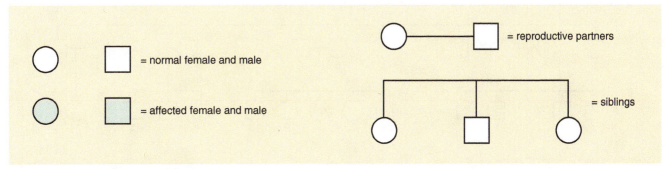

Figure 11.6 Pedigree symbols.

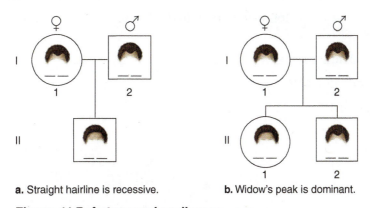

a. Straight hairline is recessive.   b. Widow's peak is dominant.

Figure 11.7 Autosomal pedigrees.

In a pedigree, Roman numerals indicate the generation, and Arabic numerals indicate particular individuals in that generation. *In Figure 11.7, you are to enter the genotype on the lines provided.* Suppose you wanted to determine the inheritance pattern for straight hairline and you knew which members of a generational family had the trait (Fig. 11.7a). The pedigree allows you to determine that straight hairline is autosomal recessive because two parents without this phenotype have a child with the phenotype. This can happen only if the parents are heterozygous and straight hairline is recessive. Similarly, a pedigree allows you to determine that widow's peak is autosomal dominant (Fig. 11.7b): A child with this phenotype has at least one parent with the dominant phenotype, but again, heterozygous parents can produce a child without widow's peak. *Give each person in Figure 11.7a and b a genotype.*

Later you will have an opportunity to see an X-linked recessive pedigree. An X-linked recessive phenotype occurs mainly in males, and it skips a generation because a female who inherits a recessive allele for the condition from her father may have a son with the condition.

Pedigree Analysis

For each of the following pedigrees, decide whether a trait is inherited as an autosomal dominant, autosomal recessive, or X-linked recessive. Then you are asked to decide the genotype of particular individuals in the pedigree.

1. Study the following pedigree:

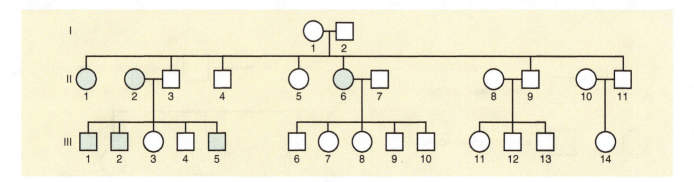

   a. Notice that neither of the original parents is affected, but several children are affected. This could

     happen only if the trait were _____.

   b. What is the genotype of the following individuals? Use *A* for the dominant allele and *a* for the recessive allele.

     Generation I, individual 1: _____

     Generation II, individual 1: _____

     Generation III, individual 8: _____

2. Study the following pedigree:

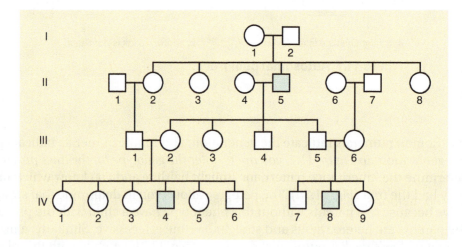

   a. Notice that only males are affected. This could happen only if the trait were _____.

   b. What is the genotype of the following individuals?

     Generation I, individual 1: _____

     Generation II, individual 8: _____

     Generation III, individual 1: _____

Construction of a Pedigree

You are a genetic counselor who has been given the following information, from which you will construct a pedigree.

1. Your data: <u>Henry</u> has a double row of eyelashes, which is a dominant trait. Both his <u>maternal grandfather</u> and his <u>mother</u> have double eyelashes. <u>Their spouses</u> are normal. Henry is married to <u>Isabella</u> and their <u>first child, Polly</u>, has normal eyelashes. The couple wants to know the chances of any child having a double row of eyelashes.

2. Choose a key for this trait.

   Key: _____ normal eyelashes   _____ double row of eyelashes

3. *Construct two pedigrees with symbols only for the underlined persons in step 1.* The pedigrees start with the maternal grandfather and grandmother and end with Polly.

   Pedigree 1   **Pedigree 2**

4. Use the pedigrees you constructed to determine the pattern of inheritance. Pedigree 1: Try out a pattern of autosomal dominant inheritance by assigning appropriate genotypes for an autosomal dominant pattern of inheritance to each person in this pedigree. Pedigree 2: Try out a pattern of X-linked dominant inheritance by assigning appropriate genotypes for this pattern of inheritance to each person in your pedigree. Which pattern is correct? _____

5. Use correct genotypes to show a cross between Henry and Isabella and, from your experience with crosses, state the expected phenotypic ratio among the offspring:

Cross	Henry	Isabella	**Phenotypic ratio:**
_____	×	_____	_____

6. What are the percentage chances that Henry and Isabella will have a child with double eyelashes? _____ "Chance has no memory" and each child has the same chance for double eyelashes. Explain why.

   _____

   _____

1. If an individual exhibits the dominant trait, do you know the genotype? Why or why not?

   _____

   _____

2. Isabella's father does not have freckles, but Mary does. What genotypes could Mary's mother have?

   _____

3. What are the chances two individuals with an autosomal recessive trait will have a child with this trait?

   _____

4. Show a cross that would produce a phenotypic ratio of 1:1 among the offspring. _____

   _____

5. If the parents are heterozygous for cystic fibrosis, what are the chances of a child having cystic fibrosis?

   _____

   _____

6. Tom has blood type AB. Show all possible genotypes for this type blood. _____

   _____

7. Mary has blood type A and Don has blood type B; can they be the parents of a child with type O blood? Show why or why not. _____

   _____

8. What syndrome is inherited when an egg carrying two X chromosomes is fertilized by a sperm carrying one Y chromosome? _____

   _____

9. What is the inheritance pattern in a pedigree if the parents are not affected and a child is affected? Give a genotype for all persons. _____

   _____

10. If only males are affected in a pedigree, what is the likely inheritance pattern for the trait? _____
    Draw a three-generation pedigree showing the inheritance of the trait from an affected grandfather to an affected grandson. (No spouses are affected.)

12

Evidences of Evolution

Learning Outcomes

12.1 Evidence from the Fossil Record
- Use the geologic timescale to trace the evolution of life in broad outline.
- Describe several types of fossils and explain how fossils help establish the sequence in the evolution of life.
- Explain how scientists use fossils to establish that organisms are related by common descent.

12.2 Evidence from Comparative Anatomy
- Explain how comparative anatomy provides evidence that humans are related to other groups of vertebrates and other groups of primates.
- Compare the human skeleton with the chimpanzee skeleton, and explain the differences on the basis of their different ways of life.
- Compare hominid skulls and hypothesize a possible evolutionary sequence.

12.3 Molecular Evidence
- Explain how molecular evidence aids the study of how humans are related to all other groups of organisms on Earth.
- Explain how molecular evidence also helps show the degree to which humans are related to other vertebrates and primates.

Introduction

Evolution is the process by which organisms are related by **common descent:** All organisms can trace their ancestry to the first cells. The process of evolution is amazingly simple: A group of organisms change over time because the members of a group most suited to the natural environment have more offspring than the others in the group. So, for example, among bacteria, those that can withstand an antibiotic leave more offspring, and with time the entire group of bacteria becomes resistant to the antibiotic. Reproduction and therefore evolution have been going on since the first cells appeared on Earth, and through studying (1) the fossil record; (2) comparative anatomy, both anatomical and embryological; and (3) molecular evidence, science is able to show that all organisms are related to one another.

    Fossils are the remains or evidence of some organism that lived long ago. Fossils can be used to trace the history of life on Earth. A comparative study of the anatomy of different groups of organisms from insects to vertebrates has shown that each group has structures of similar construction called **homologous structures.** For example, all vertebrate animals have essentially the same type of skeleton. Homologous structures signify relatedness through evolution because they can be traced to a **common ancestor** of the group. Living organisms use the same basic molecules including ATP, the carrier of energy in cells; DNA, which makes up genes; and proteins, such as enzymes and antibodies. In this laboratory you will use an antigen-antibody reaction to show the degree of evolutionary relatedness between different vertebrates.

12.1 Evidence from the Fossil Record

The geologic timescale, which was developed by both geologists and paleontologists, depicts the history of life based on the fossil record (Table 12.1). A **fossil** is any evidence of the existence of an organism in ancient times as opposed to modern times. Paleontologists specialize in removing fossils from the Earth's crust. In this section, we will study the geologic timescale and then examine some fossils.

Geologic Timescale

Divisions of the Timescale

Notice that the timescale divides the history of Earth into eras, then periods, and then epochs. The four eras span the greatest amounts of time, and the epochs are the shortest time frames. Notice that only the periods of the Cenozoic era are divided into epochs, meaning that more attention is given to the evolution of primates and flowering plants than to the earlier evolving organisms. List the four eras in the timescale,

starting with Precambrian time: _____

1. Using the geologic timescale, you can trace the history of life by beginning with Precambrian time at the bottom of the timescale. The timescale indicates that the first cells (the prokaryotes) arose some 3,500 MYA. The prokaryotes evolved before any other group. Why do you read the timescale starting at

   the bottom? _____

2. The Precambrian time was very long, lasting from the time the Earth first formed until 542 MYA. The fossil record during the Precambrian time is meager, but the fossil record from the Cambrian period onward is rich (for reasons still being determined). This helps explain why the timescale usually does not show any periods until the Cambrian period of the Paleozoic era. You can also use the timescale to check when certain groups evolved and/or flourished.
   Example: During the Ordovician period, the nonvascular plants appear on land, and the first jawless and jawed fishes appear in the seas.

   During the _____ era and the _____ period, the first flowering plants appear. How many

   million years ago was this? _____

3. On the timescale, note the Carboniferous period. During this period great swamp forests covered the land. These are also called coal-forming forests because, with time, they became the coal we burn today. How do you know that the plants in this forest were not flowering trees, as most of our trees are

   today? _____

   What type of animal was diversifying at this time? _____

4. You should associate the Cenozoic era with the evolution of humans. Among mammals, humans are

   primates. During what period and epoch did primates appear? _____

   Among primates, humans are hominids. During what period and epoch did hominids appear? _____

   _____ The scientific name for humans is *Homo sapiens*.

   What period and epoch is the age of *Homo sapiens*? _____

Table 12.1 The Geologic Timescale: Major Divisions of Geological Time with Some of the Major Evolutionary Events of Each Geological Period

Era	Period	Epoch	MYA	Plant and Animal Life
Cenozoic*	Neogene	Holocene	0–0.01	AGE OF HUMAN CIVILIZATION; destruction of tropical rain forests accelerates extinctions.
				SIGNIFICANT MAMMALIAN EXTINCTION
		Pleistocene	0.01–2	Modern humans appear; modern plants spread and diversify.
		Pliocene	2–6	First hominids appear; modern angiosperms flourish.
		Miocene	6–24	Apelike mammals, grazing mammals, and insects flourish; grasslands spread; forests contract.
	Paleogene	Oligocene	24–37	Monkeylike primates appear; modern angiosperms appear.
		Eocene	37–58	All modern orders of mammals are present; subtropical forests flourish.
		Paleocene	58–65	Primates, herbivores, carnivores, insectivores are present; angiosperms diversify
				MASS EXTINCTION: DINOSAURS AND MOST REPTILES
Mesozoic	Cretaceous		65–144	Placental mammals and modern insects appear; angiosperms spread and conifers persist.
	Jurassic		144–208	Dinosaurs flourish; birds and angiosperms appear.
				MASS EXTINCTION
	Triassic		208–250	First mammals and dinosaurs appear; forests of conifers and cycads dominate land; corals and molluscs dominate seas.
				MASS EXTINCTION
Paleozoic	Permian		250–286	Reptiles diversify; amphibians decline; gymnosperms diversify.
	Carboniferous		286–360	Amphibians diversify; reptiles appear; insects diversify. Age of great coal-forming forests.
				MASS EXTINCTION
	Devonian		360–408	Jawed fishes diversify; insects and amphibians appear; seedless vascular plants diversify and seed plants appear.
	Silurian		408–438	First jawed fishes and seedless vascular plants appear.
				MASS EXTINCTION
	Ordovician		438–510	Invertebrates spread and diversify; jawless fishes appear; nonvascular plants appear on land.
	Cambrian		510–543	Marine invertebrates with skeletons are dominant and invade land, and marine algae flourish.
Precambrian time			600	Oldest soft-bodied invertebrate fossils
			1,400–700	Protists evolve and diversify.
			2,000	Oldest eukaryotic fossils
			2,500	O_2 accumulates in atmosphere.
			3,500	Oldest known fossils (prokaryotes)
			4,500	Earth forms.

* Many authorities divide the Cenozoic era into the Tertiary period (contains Paleocene, Eocene, Oligocene, Miocene, and Plicene) and the Quaternary period (contains Pleistocene and Holocene).

Dating Within the Timescale

The timescale provides both relative dates and absolute dates. When you say, for example, "Flowering plants evolved during the Jurassic period," you are using relative time, because flowering plants evolved earlier or later than groups in other periods. If you use the dates that are given in millions of years (MYA), you are using absolute time. Absolute dates are usually obtained by measuring the amount of a radioactive isotope in the rocks surrounding the fossils. Why wouldn't you expect to find human fossils and dinosaur fossils together in rocks dated similarly? _____

Limitations of the Timescale

Because the timescale tells when various groups evolved and flourished, it might seem that evolution has been a series of events leading only from the first cells to humans. This is not the case; for example, prokaryotes (bacteria and archaea) never declined and are still the most abundant and successful organisms on Earth. Even today, they constitute up to 90% of the total weight of organisms.

Then, too, the timescale lists mass extinctions, but it doesn't tell when specific groups became extinct. **Extinction** is the total disappearance of a species or a higher group; **mass extinction** occurs when a large number of species disappear in a few million years or less. For lack of space, the geologic timescale can't depict in detail what happened to the members of every group mentioned. Figure 12.1 does show how mass extinction affected a few groups of animals. Which of the animals shown in Figure 12.1 suffered the most during the **P-T extinction** (Permian-Triassic extinction)? _____

The **K-T extinction** occurred between the Cretaceous and the Tertiary periods. Which animals shown in Figure 12.1 became extinct during the K-T extinction? _____

Figure 12.1 shows only periods and no eras. *Fill in the eras on the lines provided in the figure.*

Figure 12.1   Mass extinctions.
Five significant mass extinctions and their effects on the abundance of certain forms of marine and terrestrial life. The width of the horizontal bars indicates the varying number of each life-form considered.

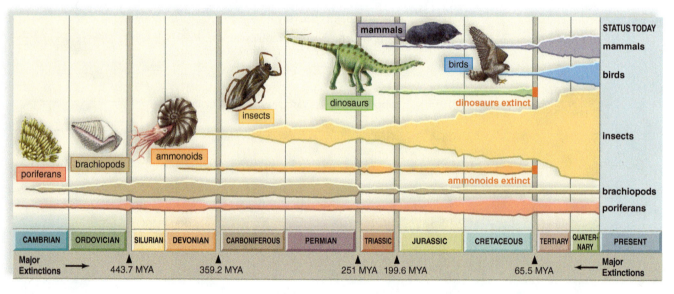

Era   _____   _____   _____

Fossils

The fossil record depends heavily on anatomical data to show the evolutionary changes that have occurred as one group gives rise to another group. Why would that be? _____

_____

Invertebrate Fossils

The **invertebrates** are animals without a backbone. Those with a hard shell, such as certain molluscs (e.g., clams), arthropods (e.g., crabs and insects), and echinoderms (e.g., sea stars), make good fossils.

Observation: Invertebrate Fossils

1. Obtain a box of selected fossils. If the fossils are embedded in rocks, examine the rock until you have found the fossil. Fossils are embedded in rocks because the sediment that originally surrounded them hardened over time. Most fossils consist of hard parts, such as shells, bones, or teeth, because these parts are not consumed or decomposed over time. One possible reason the Cambrian might be rich in

   fossils is that organisms now had _____ , whereas before they did not.

2. The kit you are using, or your instructor, will identify which of the fossils are invertebrate animals. List the names of these fossils in Table 12.2 and give a description of the fossil. Use the geologic timescale (Table 12.1) to tell the era and period they were most abundant; list the fossils from the latest (top) to earliest (bottom).

Table 12.2 Invertebrate Fossils		
Type of Fossil	Era, Period	Description

Observation: Vertebrate Fossils

The various groups of vertebrates are shown in Table 12.1. Today it is generally agreed that birds are reptiles rather than being a separate group. That means that the major groups of vertebrates are (1) various types of fishes (jawless, jawed, cartilaginous, and bony fishes), (2) amphibians, such as frogs and salamanders, (3) reptiles, such as lizards and crocodiles, and (4) mammals. There are many types of mammals, from whales to mice to humans. The fossil record can rely on skeletal differences, such as limb structure, to tell if an animal is a mammal.

Which of the fossils available to you are vertebrates? _____

_____

Use Table 12.1 to associate each fossil with the particular era and period when this type animal was most abundant. Fill in Table 12.3 according to sequence of the time frames from the latest (top) to earliest (bottom).

Table 12.3 Vertebrate Fossils

Type of Fossil	Era, Period	Description

Observation: Plant Fossils

See Figure 12.2, which shows the main groups of plants. The fossil record for plants is not as good as that for invertebrates and vertebrates because plants have no hard parts that are easily fossilized.

Figure 12.2   Plant groups.   mosses   ferns   gymnosperms   angiosperms

Plants that have no hard parts become fossils when their impressions are filled in by minerals. Use Table 12.1 to associate each fossilized plant in your kit with a particular era and period. Assume trees are flowering plants and associate them with the era and period when flowering plants were most abundant. Fill in Table 12.4 according to sequence from the latest (top) to earliest (bottom).

Table 12.4   Plant Fossils

Type of Fossil	Era, Period	Description of Fossil

Summary of Evidence from the Fossil Record

In this section, you studied the geologic timescale and various fossils represented in the record. The geologic timescale gives powerful evidence of evolution because

1. Fossils are _____.

2. Younger fossils and not older fossils are more like _____.

3. In short, the fossil record shows that _____.

12.2   Evidence from Comparative Anatomy

In the study of evolutionary relationships, parts of organisms are said to be **homologous** if they exhibit similar basic structures and embryonic origins (Fig. 12.3). If parts of organisms are similar in function only, they are said to be **analogous.** Only homologous structures indicate an evolutionary relationship and are used to classify organisms.

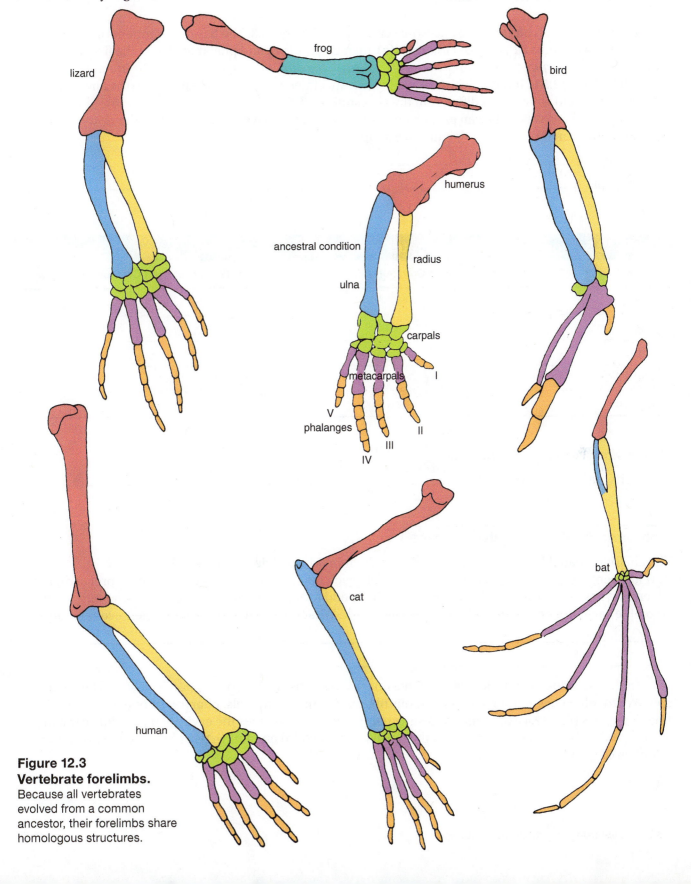

Figure 12.3
Vertebrate forelimbs.
Because all vertebrates evolved from a common ancestor, their forelimbs share homologous structures.

Comparison of Adult Vertebrate Forelimbs

The limbs of vertebrates are homologous structures (Fig. 12.3). The similarity of homologous structures is explained by descent from a common ancestor.

1. Find the forelimb bones of the ancestral vertebrate in Figure 12.3. The basic components are the humerus (h), ulna (u), radius (r), carpals (c), metacarpals (m), and phalanges (p) in the five digits.
2. *Label the corresponding forelimb bones of the lizard, the bird, the bat, the cat, and the human.*
3. Fill in Table 12.5 to indicate which bones in each specimen appear to most resemble the ancestral condition and which most differ from the ancestral condition.
4. Adaptation to a way of life can explain the modifications that have occurred. Relate the change in bone structure to mode of locomotion in two examples.

   Example 1: _____

   Example 2: _____

Table 12.5	Comparison of Vertebrate Forelimbs	
Animal	**Bones That Resemble Common Ancestor**	**Bones That Differ from Common Ancestor**
Lizard		
Bird		
Bat		
Cat		
Human		

Conclusion: Vertebrate Forelimbs

- Vertebrates are descended from a _____, but they are adapted to _____.

Comparison of Vertebrate Embryos

The anatomy shared by vertebrates extends to their embryological development. During early developmental stages, all animal embryos resemble each other closely; but as development proceeds, the different types of vertebrates take on their own shape and form. In Figure 12.4, the reptile and bird embryo resemble each other more than either resembles a fish. What does that tell you about their evolutionary relationship? _____

_____

   In the following observation, you will see that, as embryos, all vertebrates have a postanal tail, a dorsal spinal cord, **pharyngeal pouches,** and various organs. In aquatic animals, pharyngeal pouches become functional gills (Fig. 12.4). In humans, the first pair of pouches becomes the cavity of the middle ear and auditory tube, the second pair becomes the tonsils, and the third and fourth pairs become the thymus and parathyroid glands.

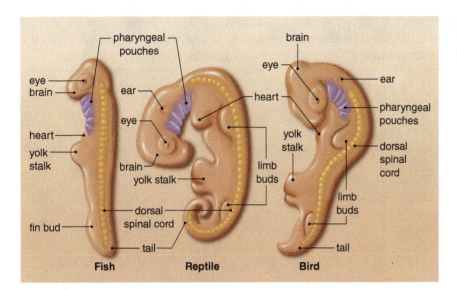

Figure 12.4 Vertebrate embryos.
The yolk stalk leads to a yolk sac which contains nourishment for the embryo.

1. Obtain prepared slides of vertebrate embryos at comparable stages of development. Observe each of the embryos using a stereomicroscope.
2. List five similarities of the embryos:

   a. _____

   b. _____

   c. _____

   d. _____

   e. _____

Conclusion: Vertebrate Embryos

Vertebrate embryos resemble one another because _____

_____.

Comparison of Chimpanzee and Human Skeletons

Chimpanzees and humans are closely related, as is apparent from an examination of their skeletons. However, they are adapted to different ways of life. Chimpanzees are adapted to living in trees and are herbivores—they eat mainly plants. Humans are adapted to walking on the ground and are omnivores—they eat both plants and meat.

Observation: Chimpanzee and Human Skeletons

Posture

Chimpanzees are arboreal and climb in trees. While on the ground, they tend to knuckle-walk, with their hands bent. Humans are terrestrial and walk erect. In Table 12.6, compare

1. **Head and torso:** Where are the head and trunk with relation to the hips and legs—thrust forward over the hips and legs or balanced over the hips and legs (see Fig. 12.5)?

Table 12.6 Comparison of Chimpanzee and Human Postures

Skeletal Part	Chimpanzee	Human
1. Head and torso		
2. Spine		
3. Pelvis		
4. Femur		
5. Knee joint		

6. Foot:
       Opposable toe
 	
       Arch		

2. **Spine:** Which animal has a long and curved lumbar (lower back) region, and which has a short and stiff lumbar region? How does this contribute to an erect posture in humans? _____

   _____

3. **Pelvis:** Chimpanzees sway when they walk because lifting one leg throws them off balance. Which animal has a narrow and long pelvis, and which has a broad and short pelvis? Record your observations in Table 12.6.

4. **Femur:** In humans, the femur better supports the trunk. In which animal is the femur angled between the pelvic girdle and the knee? In which animal is the femur straighter with less angle? Record your observations in Table 12.6.

5. **Knee joint:** In humans, the knee joint is modified to support the body's weight. In which animal is the femur larger at the bottom and the tibia larger at the top? Record your observations in Table 12.6.

6. **Foot:** In humans, the foot is adapted for walking long distances and running with less chance of injury.

   In which animal is the big toe opposable?

   _____ How does an opposable toe assist chimpanzees? _____

   _____

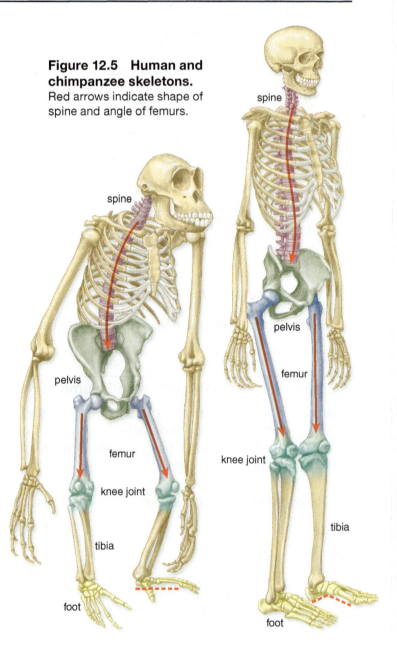

Figure 12.5   Human and chimpanzee skeletons.
Red arrows indicate shape of spine and angle of femurs.

spine

spine

pelvis

femur

knee joint

tibia

foot

pelvis

femur

knee joint

tibia

foot

Which foot has an arch? _____ How does an arch assist humans? _____

_____

Record your observations in Table 12.6.

7. In which is the neck straighter to correlate with the posture and stance of the two organisms?

_____

Skull Features

Humans are omnivorous. A diet rich in meat does not require strong grinding teeth or well-developed facial muscles. Chimpanzees are herbivores, and a vegetarian diet requires strong teeth and strong facial muscles that attach to bony projections. Compare the skulls of the chimpanzee and the human in Figure 12.6 and answer the following questions:

1. **Supraorbital ridge:** For which skull is the supraorbital ridge (the region of frontal bone just above the eye socket) thicker? Record your observations in Table 12.7.
2. **Sagittal crest:** Which skull has a sagittal crest, a projection for muscle attachments that runs along the top of the skull? Record your observation in Table 12.7.
3. **Frontal bone:** Compare the slope of the frontal bones of the chimpanzee and human skulls. How are they different? Record your observations in Table 12.7.
4. **Teeth:** Examine the teeth of the adult chimpanzee and adult human skulls. Are the incisors (two front teeth) vertical or angled? Do the canines overlap the other teeth? Are the molars larger or moderate in size? Record your observations in Table 12.7
5. **Chin:** What is the position of the mouth and chin in relation to the profile for each skull? Record your observations in Table 12.7.

Figure 12.6 Chimpanzee and human skulls.

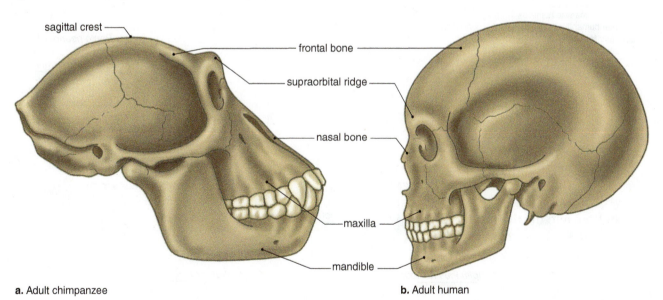

a. Adult chimpanzee   b. Adult human

Table 12.7 Skull Features of Chimpanzees and Humans

Feature	Chimpanzee	Human
1. Supraorbital ridge		
2. Sagittal crest		
3. Slope of frontal bone		
4. Teeth		
5. Chin		

Conclusions: Chimpanzee and Human Skeletons

- Do your observations show that the skeletal differences between chimpanzees and humans can be related to posture? _____ Explain. _____

- Do your observations show that diet can be related to the skull features of chimpanzees and humans? _____ Explain. _____

Comparison of Extinct Hominid Skulls

The designation *hominid* includes humans and primates that are humanlike. Paleontologists have uncovered several fossils dated from 7.5 MYA (millions of years ago) to 30,000 years BP (before present), when humans called Cro-Magnons arose that are virtually identical to modern humans *(Homo sapiens)* (Fig. 12.7).

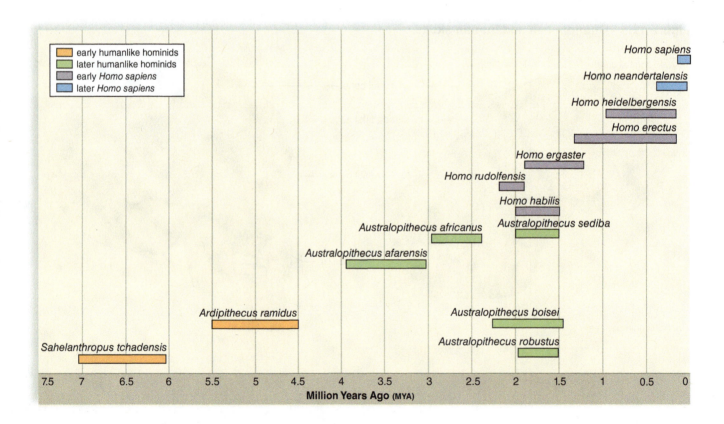

Figure 12.7 Human evolution.

Observation: Extinct Hominid Skulls

Examine the skulls of extinct hominids on display and compare them to a modern human skull in the ways noted in Tables 12.8, 12.9, and 12.10. Choose three extinct hominid skulls and record your observations in the tables.

Table 12.8 Extinct Hominid Craniums Compared to Human Cranium

Feature	Skulls		
	1.	2.	3.
a. Frontal bone (like or more flat?)			
b. Supraorbital ridge (divided or continuous?)			
c. Sagittal crest (present?)			
d. Mastoid process (flat or projecting?)			

Table 12.9 Extinct Hominid Faces Compared to Human Face

Feature	Skulls		
	1.	2.	3.
a. Nasal bones (raised or flat?)			
b. Nasal opening (larger?)			
c. Chin (projecting forward?)*			
d. Width of face (wider?)**			

*If your instructor directs you to, measure from the edge of the foramen magnum to between the incisors.

**If your instructor directs you to, measure the width of the face from mid-zygomatic arch to the other mid-arch.

Table 12.10 Extinct Hominid Dentition Compared to Human Dentition

Feature	Skulls		
	1.	2.	3.
a. Teeth rows (parallel or diverging from each other?)			
b. Incisors (vertical or angled?)			
c. Canine teeth (overlapping other teeth?)			
d. Molars (more massive?)			

Conclusions: Extinct Hominid Skulls

- Do your data appear to be consistent with the evolutionary sequence of the hominids in Figure 12.7? Explain. _____

- Report here any data you collected that would indicate a particular hominid was closer in time to humans than indicated in Figure 12.7. _____

_____

- Report here any data you collected that would indicate a particular hominid was more distant in time from humans than indicated in Figure 12.7. _____

_____

12.3 Molecular Evidence

Molecular data substantiate the comparative and developmental data that biologists have accumulated over the years. The activity of regulatory genes differ and this can account, for example, for why vertebrates have a dorsally placed nerve cord while it is ventrally placed in invertebrates. Sequencing DNA data show which organisms are closely related. Molecular data among primates is of extreme interest because it can help determine to which of the primates we are most closely related. Chromosomal and genetic data allow us to conclude that we are more closely related to chimpanzees than to other types of apes.

In this section, we note that scientists can compare the amino acid sequence in proteins to determine the degree to which any two groups of organisms are related. The sequence of amino acids in cytochrome *c,* a carrier of electrons in the electron transport chain found in mitochondria, has been determined per a variety of organisms. On the basis of the number of amino acid *differences* reported in Figure 12.8, it is concluded that the evolutionary relationship between humans and these organisms decreases in the order stated: monkeys, pigs, ducks, turtles, fishes, moths, and yeast. This conclusion agrees with the sequence of dates these organisms are found in the fossil record. Why can comparing amino acid data lead to the same conclusions as comparing DNA data?

_____

_____

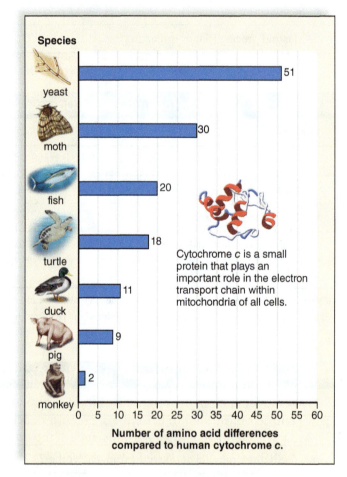

Cytochrome *c* is a small protein that plays an important role in the electron transport chain within mitochondria of all cells.

Number of amino acid differences compared to human cytochrome c.

Figure 12.8 Amino acid differences in cytochrome *c*.
The few differences found in cytochrome *c* between monkeys and humans show that, of these organisms, humans are most closely related to monkeys.

Protein Similarity Evidence

The immune system makes **antibodies** (proteins) that react with foreign proteins, termed **antigens.** Antigen-antibody reactions are specific. An antibody will react only with a particular antigen. In today's procedure, it is assumed rabbit antibodies to human antigens are in rabbit serum (Fig. 12.9). When these antibodies are allowed to react against the antigens of other animals, the stronger the antibody-antigen reaction (determined by the amount of precipitate), the more closely related the animal is to humans.

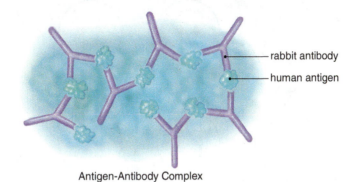

rabbit antibody

human antigen

Antigen-Antibody Complex

Figure 12.9 Antigen-antibody reaction.
When antibodies react to antigens, a precipitate appears.

Experimental Procedure: Protein Similarity Evidence

1. Obtain a chemplate (a clear glass tray with wells), one bottle of synthetic human blood serum, one bottle of synthetic rabbit blood serum, and five bottles (I to V) of blood serum test solution.
2. Put two drops of synthetic rabbit blood serum in each of the six wells in the chemplate. *Label the wells 1 to 6.* See yellow circles in Figure 12.10.
3. Add two drops of synthetic human blood serum to each well. See red circles in Figure 12.10. Stir with the plastic stirring rod that was attached to the chemplate. The rabbit serum now contains antibodies against human antigens.
4. Rinse the stirrer. (The large cavity of the chemplate may be filled with water to facilitate rinsing.)
5. Add four drops of blood serum test solution III (contains human antigens) to well 6. Describe what

   you see. _____

   _____

   This well will serve as the basis by which to compare all the other samples of test blood serum.
6. Now add four drops of blood serum test solution I to well 1. Stir and observe. Rinse the stirrer. Do the same for each of the remaining blood serum test solutions (II to V)—adding II to well 2, III to well 3, and so on. Be sure to rinse the stirrer after each use.
7. At the end of 10 and 20 minutes, record the amount of precipitate in each of the six wells in Figure 12.10. Well 6 is recorded as having + + + + amount of precipitate after both 10 and 20 minutes. Compare the other wells with this well (+ = trace amount; 0 = none). Holding the plate slightly above your head at arm's length and looking at the underside toward an overhead light source will allow you to more clearly determine the amount of precipitate.

Figure 12.10 Protein similarity.

The greater the amount of precipitate, the more closely related an animal is to humans.

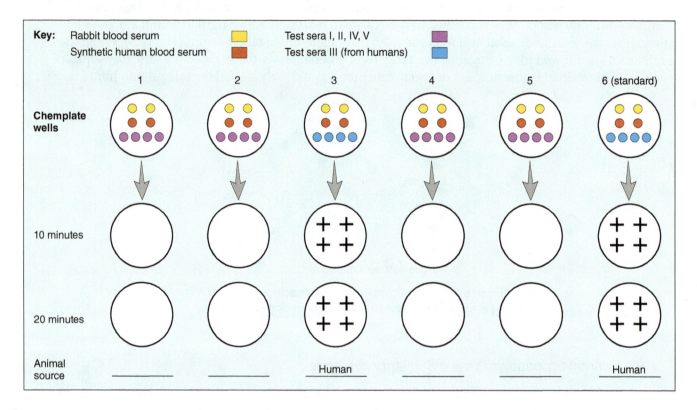

Conclusions: Protein Similarity Evidence

- The last row in Figure 12.10 tells you that the test serum in well 3 is from a human. How do your test results confirm this? _____

- Aside from humans, the test sera (supposedly) came from a pig, a monkey, an orangutan, and a chimpanzee. Which is most closely related to humans—the pig or the chimpanzee? _____

- Judging by the amount of precipitate, complete the last row in Figure 12.10 by indicating which serum you believe came from which animal. On what do you base your conclusions? _____

  _____

- In Section 12.2, a comparison of bones in vertebrate forelimbs showed that vertebrates share a common ancestor. Molecular evidence shows us that, of the vertebrates studied, _____ and _____ are most closely related. In evolution, closely related organisms share a recent common ancestor. Humans share a more recent common ancestor with chimpanzees than they do with pigs.

Laboratory Review 12

1. List three types of evidence suggesting that various types of organisms are related through common descent. _____

2. Why would you *not* expect a fossil buried millions of years ago to look exactly like a modern-day organism? _____

3. A horseshoe crab has changed little in approximately 200 million years of existence. Would you expect to find that the environment of the horseshoe crab has changed minimally? Explain. _____

   _____

4. If a characteristic is found in bacteria, fungi, pine trees, snakes, and humans, when did it most likely evolve? _____ Why? _____

   _____

5. What are homologous structures, and what do they show about relatedness? _____

   _____

6. Why do humans and chicks develop similarly to reptiles? _____

   _____

   _____

7. What do DNA mutations have to do with amino acid changes in a protein? _____

   _____

8. How can the antigen-antibody reaction help determine the degree of relatedness between species? _____

   _____

   _____

9. Using plus (+) symbols, show the amount of reaction you would expect when an antibody against human serum is tested against sera from a pig, monkey, and chimpanzee. _____

   _____

10. Define the following terms:

    a. Fossil _____

    b. Common descent _____

    c. Adaptation _____

    d. Molecular evidence _____

Essentials of Biology Website

Instructors can find lab prep information and answers to all of the laboratory questions in the Laboratory Resource Guide. *Students* can practice their knowledge with quizzes, animations, flashcards, and much more.

www.mhhe.com/maderessentials4

McGraw-Hill Access Science Website

An online encyclopedia of science and technology that provides information, including videos, that can enhance the laboratory experience.

www.accessscience.com

LEARNSMART

Evidences of Evolution

13
Microbiology

Learning Outcomes

13.1 Bacteria
- Relate the structure of a bacterium to its ability to cause disease.
- Recognize colonies of bacteria on agar plates.
- Identify the three commonly recognized shapes of bacteria.
- Recognize *Gloeocapsa*, *Oscillatoria*, and *Anabaena* as cyanobacteria.

13.2 Protists
- Identify and give examples of green, brown, red, and golden-brown algae (diatoms).
- Describe the structure and importance of diatoms and dinoflagellates.
- Distinguish protozoans on the basis of locomotion.

13.3 Fungi
- Describe the general characteristics of fungi.
- Compare sexual reproduction of black bread mold with that of a mushroom.
- Give evidence that fungi are agents of human disease.

Introduction

The history of life began with the evolution of the prokaryotic cell. Although the prokaryotic cell contains genetic material, it is not located in a nucleus, and the cell also lacks any other type of membranous organelle. At one time prokaryotes were believed to be a unified group, but based on molecular data, they are now divided into two major groups—**domain Bacteria** and **domain Archaea.** The eukaryotic cell is more closely related to the archaea than the bacteria. Eukaryotes in **domain Eukarya** have a membrane-bounded nucleus and membranous organelles. Prokaryotes resemble each other structurally but are metabolically diverse. Eukaryotes, on the other hand, are structurally diverse and exist as protists, fungi, plants, and animals. This lab pertains to these organisms.

	Domain	Cell Structure	Nutrition
Bacteria	Bacteria	Unicellular	Most heterotrophic
Protists Algae	Eukarya	Unicellular, colonial, filamentous, or multicellular	Photosynthetic
Protozoans	Eukarya	Unicellular	Heterotrophic
Fungi	Eukarya	Multicellular	Heterotrophic

13.1 Bacteria

In this laboratory, you will first relate the general structure of a bacterium to its ability to cause disease. The specific shape, growth habit, and staining characteristics of bacteria are often used to identify them. Therefore, you will observe a variety of bacteria using the microscope. Aside from their medical importance, bacteria are essential in ecosystems because, along with fungi, they are decomposers that break down dead organic remains, and thereby return inorganic nutrients to plants.

Pathogenic Bacteria

Pathogenic bacteria are infectious agents that cause disease. Infectious bacteria are able to invade and multiply within a host. Some also produce a toxin. Antibiotic therapy is often an effective treatment against a bacterial infection.

We will explore how it is possible to relate the structure of a bacterium to its ability to be invasive and avoid destruction by the immune system. We will also consider what morphophysiological attributes allow bacteria to be resistant to antibiotics and to pass the necessary genes on to other bacteria.

Observation: Structure of a Bacterium

1. Study the generalized structure of a bacterium in Figure 13.1 and, if available, examine a model or view a video of a bacterium.
2. Identify the following:

   a. **capsule:** A gel-like coating outside the cell wall. Capsules often allow bacteria to stick to surfaces such as teeth. They also prevent phagocytic white blood cells from taking them up and destroying them.

   b. **pili:** Hairlike bristles that allow adhesion to surfaces. This can be how a bacterium clings to and gains access to the body prior to an infection.

   c. **conjugation pilus:** An elongated, hollow appendage used to transfer DNA to other cells. Genes that allow bacteria to be resistant to antibiotics can be passed in this manner.

   d. **flagellum:** A rotating filament that pushes the cell forward.

   e. **cell wall:** A structure that provides support and shapes the cell. The cell wall contains a substance called **peptidoglycan.** Antibiotics that prevent the formation of a cell wall are most effective against Gram-positive rather than Gram-negative bacteria. Gram-positive bacteria have a thick cell wall that stains purple with the Gram stain, and Gram-negative bacteria have a thin cell wall that stains red with the Gram stain.

   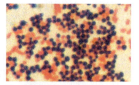
   Gram stain results

   f. **plasma membrane:** A sheet that surrounds the cytoplasm and regulates entrance and exit of molecules. Resistance to antibiotics can be due to plasma membrane alterations that do not allow the drug to bind to the membrane or cross the membrane, or to a plasma membrane that increases the elimination of the drug from the bacteria.

   g. **ribosomes:** Site of protein synthesis. Some bacteria possess antibiotic-inactivating enzymes that make them resistant to antibiotics.

   h. **nucleoid:** The location of the bacterial chromosome.

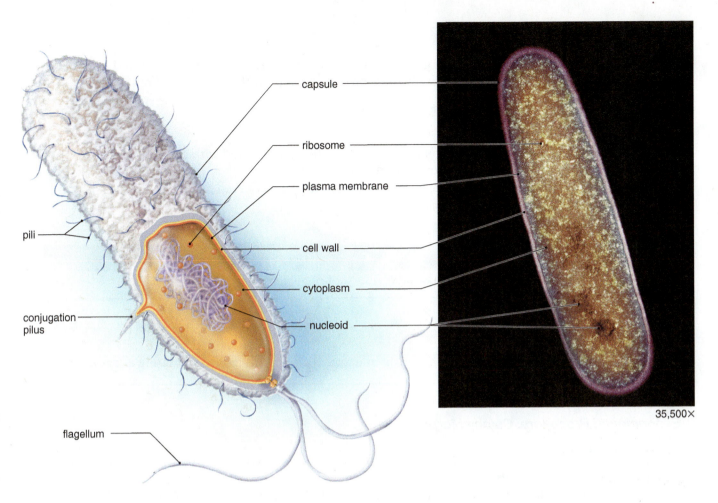

Figure 13.1 Generalized structure of a bacterium.

Also, some bacteria contain plasmids, small rings of DNA that replicate independently of the chromosomes and can be passed to other bacteria. Genes that allow bacteria to be resistant to antibiotics are often located in a plasmid.

Conclusions: Structure of a Bacterium

- Which portions of a bacterial cell aid the ability of a bacterium to cause infections?

_____

- Which portions of a bacterial cell aid the ability of a bacterium to be resistant to antibiotics?

_____

Colony Morphology

On a nutrient material called **agar**, bacteria grow as colonies. A **colony** contains cells descended from one original cell. Sometimes, it is possible to identify the type of bacterium by the appearance of the colony.

1. Do not remove the covers and view agar plates that have been inoculated with bacteria and then incubated. Notice the "colonies" of bacteria growing on the plates.
2. Compare the colonies' color, surface, and margin, and note your observations in Table 13.1. It is not necessary to identify the type of bacteria.

Table 13.1 Agar Plates	
Plate Number	Description of Colonies

Experimental Procedure: Colony Morphology

1. If available, obtain a sterile agar plate, and inoculate the plate with your thumbprint, or use a swab and inoculate the plate with material from around your teeth or inside your nose. Put your name on the plate, and place it where directed by your instructor. Remember to view the plate next laboratory period and describe what you see. _____

   _____

2. If available, obtain a sterile agar plate, and expose it briefly (at most for 10 minutes) anywhere you choose, such as in the library, your room, or your car. Describe the appearance of your plate after exposure. _____

   _____

Conclusions

- What have you discovered from this exercise? _____

   _____

Shape of Bacterial Cell

Most bacteria are found in three basic shapes: **spirillum** (spiral or helical), **bacillus** (rod), and **coccus** (round or spherical) (Fig. 13.2). Bacilli may form long filaments, and cocci may form clusters or chains. Some bacteria form endospores. An **endospore** contains a copy of the genetic material encased by heavy, protective spore coats. Spores survive unfavorable conditions and germinate to form vegetative cells when conditions improve.

Observation: Shape of Bacterial Cell

1. View the microscope slides of bacteria on display. What magnification is required to view bacteria?

    _____

2. Using Figure 13.2 as a guide, identify the three different shapes of bacteria. _____

3. Do any of the slides on display show bacterial cells with endospores? _____

    What is an endospore, and why does it have survival value? _____

    _____

Figure 13.2 Diversity of bacteria.
a. Spirillum, a spiral-shaped bacterium. **b.** Bacilli, rod-shaped bacteria. **c.** Cocci, round bacteria.

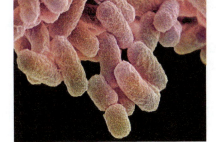

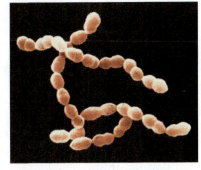

a. Spirillum: TEM 4,000×
Treponema pallidum

b. Bacilli: SEM 16,000×
Escherichia coli

c. Cocci: SEM 6,250×
Streptococcus thermophilus

Cyanobacteria

Cyanobacteria were formerly called blue-green algae because their general growth habit and appearance through a compound light microscope are similar to green algae. Electron microscopic study of cyanobacteria, however, revealed that they are structurally similar to other bacteria, particularly other photosynthetic bacteria. Although cyanobacteria do not have chloroplasts, they do have thylakoid membranes, where photosynthesis occurs.

Gloeocapsa

1. Prepare a wet mount of a *Gloeocapsa* culture, if available, or examine a prepared slide, using high power (45×) or oil immersion (if available). The single cells adhere together because each is surrounded by a sticky, gelatinous sheath (Fig. 13.3).

2. What is the estimated size of a single cell? _____

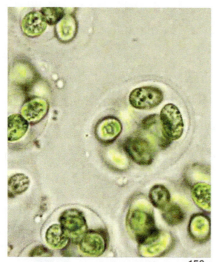
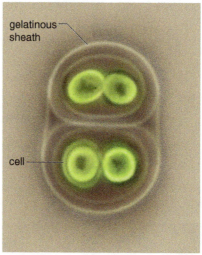

a. Micrograph at low magnification   150×

b. Micrograph at high magnification   500×

gelatinous sheath

cell

Figure 13.3   *Gloeocapsa.*

Oscillatoria

1. Prepare a wet mount of an *Oscillatoria* culture, if available, or examine a prepared slide, using high power (45×) or oil immersion (if available). This is a filamentous cyanobacterium with individual cells that resemble a stack of pennies (Fig. 13.4).

2. *Oscillatoria* takes its name from the characteristic oscillations that you may be able to see if your sample is alive. If you have a living culture, are oscillations visible? _____

Figure 13.4   *Oscillatoria.*

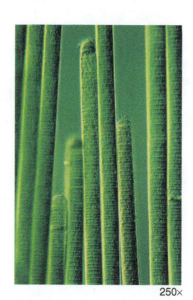

250×

Anabaena

1. Prepare a wet mount of an *Anabaena* culture, if available, or examine a prepared slide, using high power (45×) or oil immersion (if available). This is also a filamentous cyanobacterium, although its individual cells are barrel-shaped (Fig. 13.5).

2. Note the thin nature of this strand. If you have a living culture, what is its color? _____

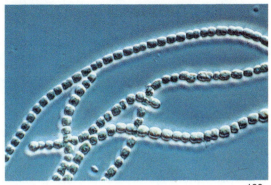

160×

Figure 13.5   *Anabaena*.

13.2   Protists

Protists were the first eukaryotes to evolve. Their diversity and complexity make it difficult to categorize them. However, the diagram in Figure 13.6 may be helpful.

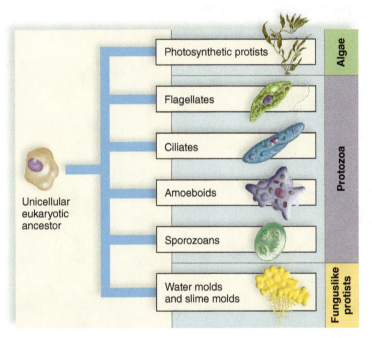

Figure 13.6   Major groups of protists.
Because the precise evolutionary relationships between these groups are not yet known, they are grouped here by major shared characteristics.

Algae

Algae is a term that is used for aquatic organisms that photosynthesize in the same manner as land plants. All photosynthetic protists contain green chlorophyll, but they also may contain other pigments that mask the chlorophyll color, and this accounts for their common names—the green algae, red algae, brown algae, and golden-brown algae.

If available, view a video showing the many forms of green algae. Notice that green algae can be single cells, filaments, colonies, or multicellular sheets. You will examine a filamentous form *(Spirogyra)* and a colonial form *(Volvox)*. A **colony** is a loose association of cells.

Spirogyra

1. Make a wet mount of live *Spirogyra,* or observe a prepared slide. *Spirogyra* is a filamentous alga, lives in fresh water, and often is seen as a green scum on the surface of ponds and lakes. The most prominent feature of the cells is the spiral, ribbonlike chloroplast (Fig. 13.7*a*). How do you think *Spirogyra* got its name? _____

_____

Spirogyra's chloroplast contains a number of circular bodies, the **pyrenoids,** centers of starch polymerization. The nucleus is in the center of the cell, anchored by cytoplasmic strands.

2. Your slide may show **conjugation,** a sexual means of reproduction illustrated in Figure 13.7*b*. If it does not, obtain a slide that does show this process. Conjugation tubes form between two adjacent filaments, and the contents of one set of cells enter the other set. As the nuclei fuse, a zygote is formed. The zygote overwinters, and in the spring, meiosis and, subsequently, germination occur. The resulting adult protist is therefore haploid.

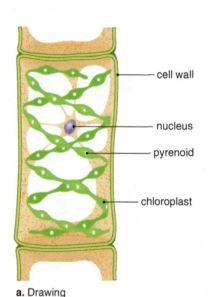

cell wall

nucleus

pyrenoid

chloroplast

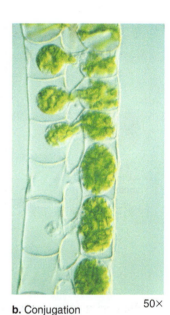

50×

Figure 13.7   *Spirogyra.*
a. *Spirogyra* is a filamentous green alga, in which each cell has a ribbonlike chloroplast. **b.** During conjugation, the cell contents of one filament enter the cells of another filament. Zygote formation follows.

a. Drawing

b. Conjugation

Volvox

1. Using a depression slide, prepare a wet mount of live *Volvox,* or study a prepared slide. *Volvox* is a green algal colony. It is motile (capable of locomotion) because the thousands of cells that make up the colony have flagella. These cells are connected by delicate cytoplasmic extensions (Fig. 13.8).

Volvox is capable of both asexual and sexual reproduction. Certain cells of the adult colony can divide to produce **daughter colonies** (Fig. 13.8) that reside for a time within the parental colony. A daughter colony esc apes the parental colony by releasing an enzyme that dissolves away a portion of the matrix of the parental colony. During sexual reproduction, some colonies of *Volvox* have cells that produce sperm, and others have cells that produce eggs. The resulting zygote undergoes meiosis and the adult *Volvox* is haploid.

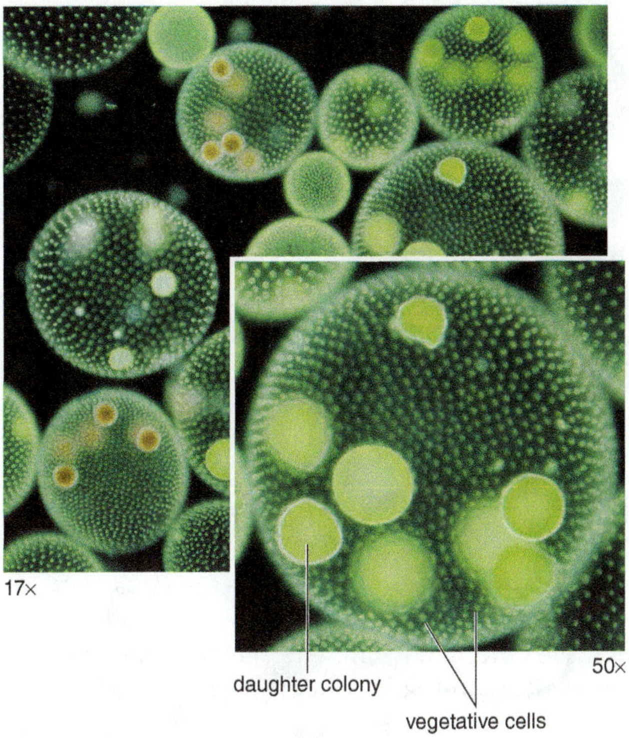

Figure 13.8 *Volvox*.
Volvox is a colonial green alga. The adult *Volvox* colony often contains daughter colonies, asexually produced by special cells.

17×

50×

daughter colony

vegetative cells

2. In Table 13.2, list the genus names of each of the green algae specimens available, and give a brief description.

Table 13.2	Green Algae Diversity
Specimen	**Description**

Observation: Brown Algae and Red Algae

Brown algae and red algae are commonly called *seaweed,* along with the multicellular green algae.

1. If available, study preserved specimens of brown algae (Fig. 13.9) which have brown pigments that mask chlorophyll's green color. These algae are large and have specialized parts. *Laminaria* algae are called **kelps.** *Fucus* is called rockweed because it is seen attached to rocks at the seashore when the tide is out (Fig. 13.9). If available, view a preserved specimen. Note the dichotomously branched body plan, so called because the **stipe** repeatedly divides into two branches (Fig. 13.9). Note also the **holdfast** by which the alga anchors itself to the rock; the **air vesicles,** or bladders, that help hold the thallus erect in the water; and the **receptacles,** or swollen tips. The receptacles are covered by small, raised areas, each with a hole in the center. These areas are cavities in which the sex organs are located, with the gametes escaping to the outside through the holes. *Fucus* is unique among algae in that as an adult it is diploid (2n) and always reproduces sexually.

Figure 13.9 Brown algae.
Laminaria and *Fucus* are seaweeds known
as kelps. They live along rocky coasts of the
north temperate zone. The other brown algae
featured, *Macrocystis* and *Nereocystis*, form
spectacular underwater "forests" at sea.

air
bladder

blade

stipe

holdfast

Fucus

Laminaria

Nereocystis

Macrocystis

2. If available study preserved specimens of red algae. Like most brown algae, the red algae are
   multicellular, but they occur chiefly in warmer seawater, growing both in shallow waters and as deep
   as light penetrates. Some forms of red algae are filamentous, but more often, they are complexly
   branched with a feathery, flat, and expanded or ribbonlike appearance. Coralline algae are red algae
   that have cell walls impregnated with calcium carbonate ($CaCO_3$).

3. In Table 13.3, list the genus names of each of the brown and red algae specimens available, and give a
   brief description.

Table 13.3	Brown and Red Algae	
Specimen	**Genus**	**Description**
1		
2		
3		
4		

Diatoms (golden-brown algae) have a yellow-brown pigment that, in addition to chlorophyll, gives them their color.

1. Make a wet mount of live diatoms, or view a prepared slide (Fig. 13.10). Describe what you see:_____

_____

_____

_____

_____

The cell wall of diatoms is in two sections, with the larger one fitting over the smaller as a lid fits over a box. Since the cell wall is impregnated with silica, diatoms are said to "live in glass houses." The glass cell walls of diatoms do not decompose, so they accumulate in thick layers subsequently mined as diatomaceous earth and used in filters and as a natural insecticide. Diatoms, being photosynthetic and extremely abundant, are important food sources for the small heterotrophs (organisms that must acquire food from external sources) in both marine and freshwater environments.

2. Prepare a wet mount of live dinoflagellates or view a prepared slide (Fig. 13.11). Describe what you see.

_____

_____

_____

_____

Dinoflagellates are photosynthetic, but they have two flagella; one is free, but the other is located in a transverse groove that encircles the animal. The beating of these flagella causes the organism to spin like a top. The cell wall, when present, is frequently divided into closely joined polygonal plates of cellulose. At times there are so many of these organisms in the ocean that they cause a condition called "red tide." The toxins given off in these red tides cause widespread fish kills and can cause paralysis in humans who eat shellfishes that have fed on the dinoflagellates.

130×

Figure 13.10 Diatoms.
Diatoms, photosynthetic protists of the oceans.

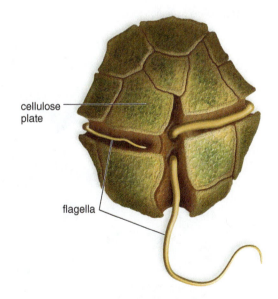

cellulose plate

flagella

Figure 13.11 Dinoflagellates.
Dinoflagellates such as *Gonyaulax* have cellulose plates.

Protozoans

The term *protozoan* refers to unicellular eukaryotes and is often restricted to heterotrophic organisms that ingest food by forming **food vacuoles.** Other vacuoles, such as **contractile vacuoles** that rid the cell of excess water, are also typical. Usually protozoans have some form of locomotion; some, such as amoebas, use **pseudopods** for locomotion and feeding; some, such as paramecia, move by **cilia;** and some, such as trypanosomes, use **flagella** (Fig. 13.12). Trypanosomes are disease-causing parasites.

   Plasmodium vivax, a common cause of malaria, is an apicomplexan, a protozoan that contains a special organelle called an apicoplast. The apicoplast assists the parasite in penetrating host cells. This type of protozoan is also often called a **sporozoan** because it goes through an asexual phase in which it exists as particulate spores. During its asexual phase, *Plasmodium vivax* lives inside red blood cells and the chills and fever of malaria occur when the red blood cells burst to release the spores. Sporozoans have no obvious means of locomotion as do the other types of protozoans illustrated in Figure 13.12. How do

sporozoans differ from other types of protozoans? _____

_____

_____

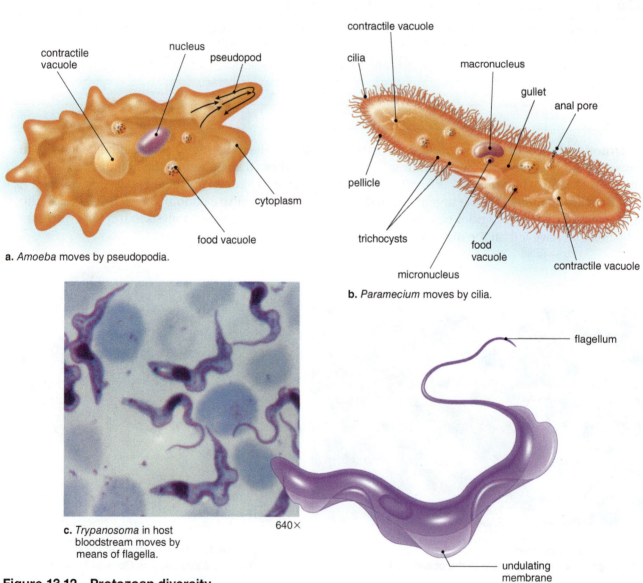

a. *Amoeba* moves by pseudopodia.

b. *Paramecium* moves by cilia.

c. *Trypanosoma* in host bloodstream moves by means of flagella.

640×

Figure 13.12   Protozoan diversity.
Protozoans are motile by the means illustrated. **a.** *Amoeba.* **b.** *Paramecium.* **c.** *Trypanosoma* in host.

1. Watch a video if available, and note the various forms of protozoans. Prepare wet mounts or examine prepared slides of protozoans as directed by your instructor. Complete Table 13.4, listing the structures for locomotion in the types of protozoans you have observed in this laboratory.

Table 13.4 Protozoans		
Name	**Structures for Locomotion**	**Observations**

2. Make a wet mount of *Euglena* by using a drop of a *Euglena* culture and adding a drop of Protoslo (methyl cellulose solution) onto the slide to slow the organisms down. Describe what you see. _____
   _____

Euglena (Fig. 13.13) typifies the problem of classifying protists. One third of all *Euglena* genera have chloroplasts; the rest do not. This discrepancy can be explained: The chloroplasts are probably green algae taken up by phagocytosis (engulfing them). A pyrenoid is a region of the chloroplast where a special type of carbohydrate is formed.

Euglena has a long flagellum that projects out of a vaselike indentation and a much shorter one that does not project out. It moves very quickly, and you will be advised to add Protoslo to your wet mount to slow it down. Like some of the protozoans discussed previously, *Euglena* is bounded by a flexible pellicle made of protein. This means it can also assume all sorts of shapes. *Euglena* lives in fresh water and contains a contractile vacuole that collects water and then contracts, ridding the body of excess water.

Figure 13.13 *Euglena.*
Euglena is a unicellular, flagellated protist.

- short flagellum
- contractile vacuole
- long flagellum
- eyespot
- photoreceptor
- carbohydrate granule
- nucleolus
- nucleus
- pellicle band
- pyrenoid
- chloroplast

a. Drawing

b. Photomicrograph   200×

3. Make a wet mount of pond water by taking a drop from the bottom of a container of pond water. Scan the slide for organisms: Start at the upper left-hand corner, and move the slide forward and back as you work across the slide from left to right. Experiment by using all available objective lenses, by focusing up and down with the fine-adjustment knob, and by adjusting the light so that it is not too bright. Identify the organisms you see by consulting Figure 13.14, and use any pictorial guides provided by your instructor.

Figure 13.14   Microorganisms found in pond water. Drawings are not actual sizes of the organisms.

Amoeba	Euglena	Paramecium	Vorticella	Euplotes	Asplanchna
Tetrahymena	Stentor	Colpoda	Didinium	Philodina	Keratella
Chilomonas	Chlamydomonas	Difflugia	Arcella	Eudorina	Pandorina
Blepharisma	Chilodonella	Stylonychia	Dileptus	Chaetonotus	Oxytricha

13.3 Fungi

Fungi (kingdom Fungi) (Fig. 13.15) are **saprotrophs** that release their digestive enzymes into the environment. Both fungi and bacteria are often referred to as "organisms of decay" because as they break down dead organic matter they release inorganic nutrients for plants. A fungal body, called a **mycelium,** is composed of many strands, called **hyphae** (Fig. 13.16). Sometimes, the nuclei within a hypha are separated by walls called septa.

Fungi produce windblown **spores** (small, haploid bodies with a protective covering) when they reproduce sexually or asexually.

Figure 13.15 Diversity of fungi.
a. Scarlet hood, an inedible mushroom. **b.** Spores exploding from a puffball. **c.** Common bread mold. **d.** Morel, an edible fungus.

Figure 13.16 Body of a fungus.
a. The body of a fungus is called a mycelium. **b.** A mycelium contains many individual chains of cells, and each chain is called a hypha.

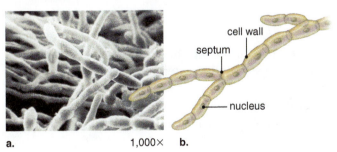

a.   1,000×   b.

Black Bread Mold

In keeping with its name, black bread mold grows on bread and any other type of bakery goods. Notice in Figure 13.17 the sporangia that produce spores in both the asexual and sexual life cycles. A **zygospore** is diploid (2n); otherwise, all structures in the asexual and sexual life cycles of bread mold are haploid (n) but following meiosis, the spores are haploid..

Figure 13.17 Black bread mold.
The yellow arrows represent sexual reproduction during which two hyphae tips fuse, and then two nuclei fuse, forming a zygote that develops a thick resistant wall becoming a zygospore. When conditions are favorable the zygospore germinates, and meiosis within a sporangium produces windblown spores. During asexual reproduction, one haploid parent produces haploid spores in sporangia.

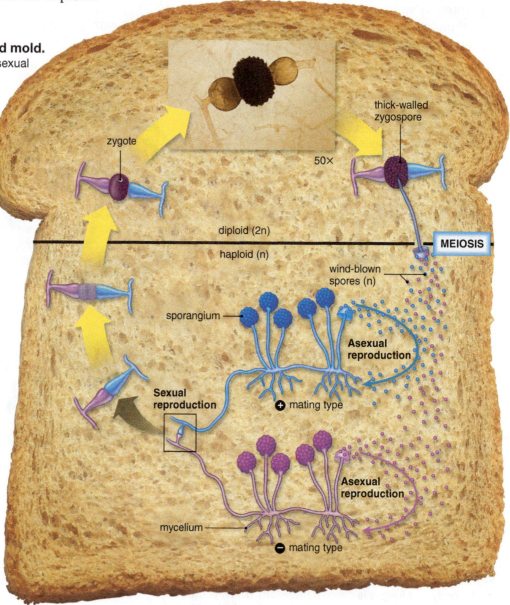

zygote

thick-walled zygospore

50×

diploid (2n)

haploid (n)

MEIOSIS

wind-blown spores (n)

sporangium

Asexual reproduction

Sexual reproduction

⊕ mating type

Asexual reproduction

mycelium

⊖ mating type

Observation: Black Bread Mold

1. If available, examine bread that has become moldy. Do you recognize black bread mold on the bread?
2. Obtain a petri dish that contains living black bread mold. Observe with a stereomicroscope. *Label the mycelium and a sporangium in Figure 13.18a.*
3. View a prepared slide of *Rhizopus,* using both a stereomicroscope and the low-power setting of a light microscope. The absence of cross walls in the hyphae is an identifying feature of zygospore fungi. *Label the mycelium and zygospore in Figure 13.18b.*

Figure 13.18   Microscope slides of black bread mold.
a. Asexual life-cycle structures. **b.** Sexual life-cycle structures.

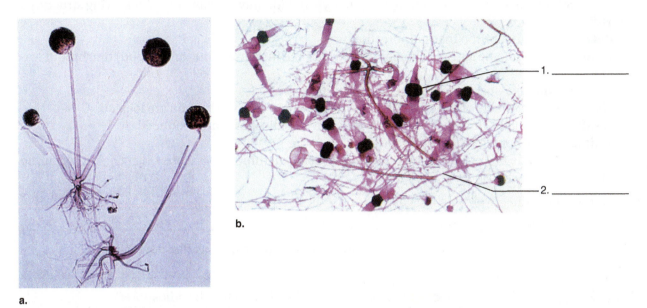

a.

b.

1. _____

2. _____

Club Fungi

Club fungi are just as familiar as black bread mold to most laypeople because they include the mushrooms. A gill mushroom consists of a stalk and a terminal cap with gills on the underside (Fig. 13.19). The stalk and cap, called a **basidiocarp,** is a fruiting body that arises following the union of + and − hyphae. The gills bear basidia, club-shaped structures where nuclei fuse, and meiosis occurs during spore production. The spores are called **basidiospores.**

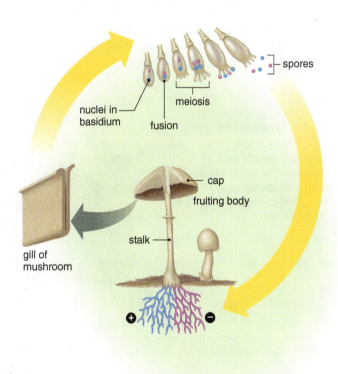

Figure 13.19   Sexual reproduction produces mushrooms.
Fusion of + and − hyphae tips results in hyphae that form the mushroom (a fruiting body). The nuclei fuse in clublike structures attached to the gills of a mushroom, and meiosis produces spores.

1. Obtain an edible mushroom—for example, *Agaricus*—and identify as many of the following structures as possible:
   a. **Stalk:** The upright portion that supports the cap.
   b. **Annulus:** A membrane surrounding the stalk where the immature (button-shaped) mushroom was attached.
   c. **Cap:** The umbrella-shaped basidiocarp of the mushroom.
   d. **Gills:** On the underside of the cap, radiating lamellae on which the basidia are located.
   e. **Basidia:** On the gills, club-shaped structures where basidiospores are produced.
   f. **Basidiospores:** Spores produced by basidia.
2. View a prepared slide of a cross section of *Coprinus*. Using all three microscope objectives, look for the gills, basidia, and basidiospores.
3. Can you see individual hyphae in the gills? _____
4. Are the basidiospores inside or outside of the basidia? _____
5. What type of nuclear division does the zygote undergo to produce the basidiospores? _____

   _____
6. Can you suggest a reason for some of the basidia having fewer than four basidiospores? _____

   _____
7. What happens to the basidiospores after they are released? _____

   _____

Fungi and Human Diseases

Fungi cause a number of human diseases. Oral thrush is a yeast infection of the mouth common in newborns and AIDS patients (Fig. 13.20*a*). Ringworm is a group of related diseases caused by the fungus *Tinea*. The fungal colony grows outward, forming a ring of inflammation (Fig. 13.20*b*). Athlete's foot is a form of *Tinea* that affects the foot, mainly causing itching and peeling of the skin between the toes (Fig. 13.20*c*).

Figure 13.20 Human fungal diseases.
a. Thrush, or oral candidiasis, is characterized by the formation of white patches on the tongue.
b. Ringworm and **c.** athlete's foot are caused by *Tinea* spp.

Laboratory Review 13

1. What role do bacteria and fungi play in ecosystems? _____
_____

2. What type of semisolid medium is used to grow bacteria? _____

3. What is the scientific name for spherical bacteria? _____

4. It is sometimes said that diatoms live in what kind of "houses"? _____

5. What type of nutrition do algae have? _____

6. Name a colonial alga studied today. _____

7. Bacteria have what substance in their cell walls? _____
_____

8. What color are Gram-negative bacteria following Gram staining? _____

9. Once called the blue-green algae, cyanobacteria are now classified as what? _____

10. What do you call the projection that allows amoeboids to move and feed? _____

11. In what type of environment are you likely to find brown algae, such as *Fucus*? _____
_____

12. The stalk and cap of a mushroom are scientifically termed what? _____

13. Describe the saprotrophic nutrition of fungi. _____
_____

14. What do fungi produce during both sexual and asexual reproduction? _____

15. Why aren't all the organisms studied today in the domain Eukarya? _____
_____

16. In general, how does sexual reproduction differ from asexual reproduction among fungi? _____
_____

14

Plant Evolution

Introduction

Your study of plant evolution in this laboratory will emphasize four groups of plants: the mosses; the ferns; the gymnosperms, represented by the pine tree; and the angiosperms, the flowering plants. While each of these groups of plants is successfully adapted to living on land, adaptation to the land environment is best demonstrated by the flowering plants (Fig. 14.1). The number and kinds of flowering plants are much greater than those of all the other groups of plants.

Adaptation to a land environment includes the ability to prevent excessive loss of water into the atmosphere; to obtain and transport water and nutrients to all parts of the plant; to support a large body against the pull of gravity; and to reproduce without dependence on external water.

With regard to human beings, consider that skin protects us from drying out, blood transports water and nutrients about the body, the skeleton supports us, males have a penis for delivering flagellated sperm to the female, and the embryo and fetus are protected from drying out within the uterus of the female.

Figure 14.1   The flowering plants.
The flowering plants provide us with much of our food, including grains, potatoes, and beans. Here, we see the flower and the fruit of a watermelon plant.

Figure 14.2   Evolution of plants.

Land plants arose from a common green algal ancestor. The evolution of land plants is marked by five significant events: protection of the embryo, evolution of vascular tissue, evolution of megaphylls, evolution of the seed, and evolution of the flower.

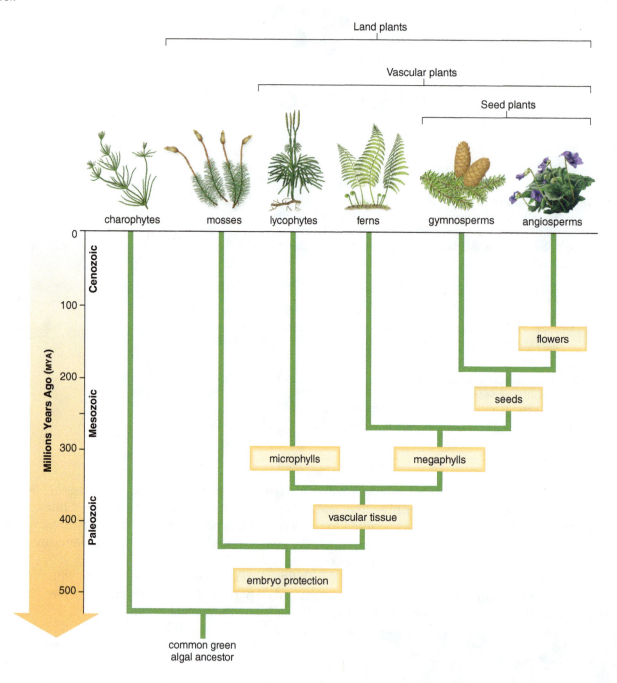

14.1 Evolution of Plants

Figure 14.2 shows the evolution of land plants. What significant evolutionary events led to adaptation of plants to a land existence? _____

_____

Algal Ancestor of Land Plants

The charophytes are green algae that share a common ancestor with land plants. This common ancestor may have resembled *Chara*, which can live in warm, shallow ponds that occasionally dry up. Adaptation to such periodic desiccation may have facilitated the ability of certain members of the common ancestor population to invade land and, with time, become the first land plants.

Observation: **Chara**

Examine a living *Chara* (Fig. 14.3). How does it superficially resemble a land plant? _____

_____

Chara is a filamentous green alga that consists of a primary branch and many side branches. Each branch has a series of very long cells. Note where one cell ends and the other begins. Measure the length of one cell. _____ Gently pick up and handle *Chara* while you are examining it. What does it feel like? _____ Its cell walls are covered with calcium carbonate deposit.

Figure 14.3 *Chara.*
Chara is an example of a stonewort, the type of green alga believed to be most closely related to the land plants.

Chara, several individuals   One individual

branch
main axis
node

Conclustions: **Chara**

• What characteristics cause *Chara* to resemble land plants? _____

_____

• Why are *Chara* called stoneworts? _____

Alternation of Generations

All plants have a life cycle known as alternation of generations (Fig. 14.4). In this life cycle, there are two mature stages, known as the sporophyte and the gametophyte. The **sporophyte,** the 2n generation, produces spores, by the process of meiosis, in structures called sporangia (sing., sporangium). A **spore** is a haploid reproductive cell that produces a new generation that is also haploid. Spores develop into the gametophyte. The **gametophyte,** the n generation, produces gametes that later fuse to form a zygote. A zygote develops into the sporophyte, completing the cycle.

In plants, one generation is **dominant,** meaning that it lasts longer and is the generation we refer to as the plant. The gametophyte is dominant in mosses, and the sporophyte is dominant in ferns, gymnosperms, and angiosperms. In plants with dominant sporophytes, the sporophyte has vascular (transport) tissue that conducts water from the roots to the leaves. Is it beneficial for a sporophyte, the generation that has vascular tissue, to be dominant? _____

_____

Why? _____

Figure 14.4   Alternation of generations.
In the life cycle of plants, called alternation of generations, the sporophyte is the 2n generation that produces spores by meiosis. The gametophyte is the n generation that produces the gametes. When the gametes fuse, a new sporophyte comes into being.

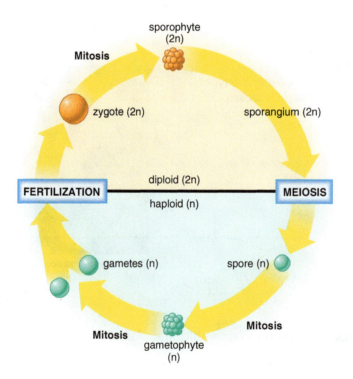

14.2 Seedless Plants

Mosses and ferns are both seedless plants. Mosses and their relatives, called the bryophytes, are low-lying plants, called the nonvascular plants because they lack vascular tissue. Ferns, characterized by large leaves, do have vascular tissue, but even so, share other characteristics with the bryophytes. For example, they are both seedless plants.

Mosses

The bryophytes (mosses and liverworts) were the first plants to live on land. The gametophyte is dominant in these nonvascular plants. The gametophyte produces eggs within archegonia and swimming sperm in antheridia (Fig. 14.5). The bryophytes are dependent on external moisture to ensure fertilization because the sperm must swim to the egg. However, the zygote developing within the archegonia is protected from drying out. The bryophytes have another adaptation to life on land in that the spores, produced by the dependent sporophyte, are windblown. The spores disperse the gametophyte.

Figure 14.5 Life cycle of the moss.
In mosses, the gametophyte consists of leafy shoots that produce flagellated sperm within antheridia and eggs within archegonia. Following fertilization, the sporophyte, consisting of a stalk and capsule (a sporangium), is dependent on the gametophyte.

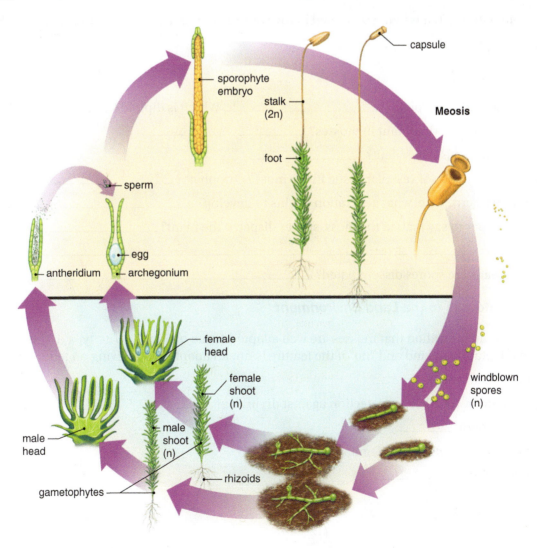

Life Cycle of Mosses

Study the life cycle of the moss (see Fig. 14.5) and find the gametophyte. Examine a living gametophyte or a plastomount of this generation. Describe its appearance. _____ _____ Considering that this is the generation we refer to as the "moss," what generation is dominant in mosses? _____ Describe the sporophyte. _____

1. Study the prepared slide of the female shoot head (top of female shoot) and the male shoot head (top of the male shoot). Find an **archegonium** in a female shoot head and locate an egg in at least one of these. Find an **antheridium** in a male shoot head. An antheridium is filled with many flagellated sperm. When sperm produced by the antheridia swim in a film of water to the eggs in the archegonia, zygotes result. A zygote develops into a new sporophyte.

2. Examine the plastomount of a shoot with a sporophyte attached. The sporophyte is dependent on the female shoot. Why female? _____

3. Examine a slide of a longitudinal section through the sporophyte of the moss. Identify the stalk and sporangium. What is being produced in the sporangium? _____
   By what process? _____

The Life Cycle of a Moss

1. Which generation is haploid? _____ Which is diploid? _____
   Which generation is dominant in mosses? _____
   Which generation is dependent? _____
2. Is there any evidence of vascular tissue in the moss sporophyte? _____
3. When spores germinate, what generation begins to develop? _____
4. Why is it proper to say that, in the moss, spores disperse the plant? _____
   _____
5. By what means are spores disseminated? _____

Adaptation of Mosses to the Land Environment

Which of these is an indication that mosses are well adapted to life on land? Write "yes" if the feature is an adaptation to land and "no" if the feature is not an adaptation to living on land.

Lack of vascular tissue _____

Body covered by a cuticle, a protection against drying out _____

Flagellated sperm _____

Egg and embryo protected by female shoot _____

Spores are windblown _____

Lycophytes

Lycophytes (phylum Lycophyta) are commonly called **club mosses.** Lycophytes are representative of the first vascular plants. The presence of vascular tissue means that lycophytes have true roots, stems, and leaves. They have an aerial stem and a horizontal root called a rhizome. A rhizome has vascular tissue but its attached thread like structures called rhizoids do not have vascular tissue. Rhizoids increase the surface area for absorption of water and minerals. The leaves of lycophytes are called **microphylls** because they are so small and have only one strand of vascular tissue.

Observation: Lycophytes

1. Examine a living or preserved specimen of *Lycopodium* (Fig. 14.6).
2. Note the shape and the size of the microphylls and the branches of the stems.
3. Note the terminal clusters of leaves, called **strobili,** that are club-shaped and bear sporangia.
4. *Label strobili, leaves, stem, and rhizoids in Figure 14.6.*
5. Examine a prepared slide of a Lycopodium that shows the sporangia with spores inside. The spore develops into a microscopic gametophyte that remains in the soil.

Figure 14.6 *Lycopodium.*
In the club moss *Lycopodium*, green photosynthetic stems are covered by scalelike leaves, and spore-bearing leaves are clustered in strobili.

Ferns

All the other plants to be studied in this laboratory are vascular plants. Ferns are seedless vascular plants in which the dominant sporophyte possesses vascular tissue. The windblown spores develop into an independent gametophyte that is water dependent because it lacks vascular tissue and because it produces flagellated sperm. The sperm must swim from the antheridia to the archegonia, where the eggs are produced. The zygote, protected from drying out within an archegonium, develops directly into the sporophyte.

Observation: Fern Life Cycle

1. Study the life cycle of the fern (Fig. 14.7) and find the sporophyte and gametophyte. This large, complexly divided sporophyte leaf is known as a **frond.** Fronds arise from an underground stem called a **rhizome.** Examine a preserved specimen of a frond and, on the underside, notice brownish **sori,** each one a cluster of many sporangia (Fig. 14.8). What is being produced in the

   sporangia? _____ Considering that it is this generation that we call

   the fern, what generation is dominant in ferns? _____

Figure 14.7   Fern life cycle.

The frond is the dominant sporophyte generation that produces windblown spores. A spore gives rise to an independent gametophyte, called a prothallus. The prothallus produces gametes. When the gametes fuse, a new sporophyte begins to develop.

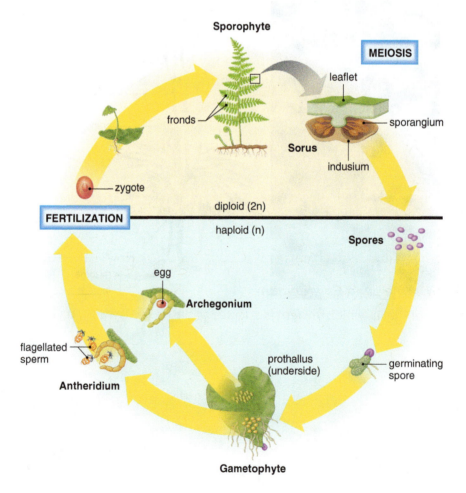

Figure 14.8 Underside of frond leaflets showing sori.

sorus

2. Examine a prepared slide of young sporangium, cross section. Study the slide carefully and locate the fern leaf and a sorus (pl., sori). Within a sorus, find the sporangia and spores. Notice the **indusium,** a shelflike structure that protects the sporangia until they are mature.

3. Examine a plastomount showing the fern life cycle and find the gametophyte generation. It is a small, heart-shaped structure called the **prothallus.** Most persons do not realize that this structure exists as a part of the fern life cycle. What is the function of the gametophyte? _____

_____

4. Examine a whole mount slide of a fern prothallium-archegonia. What is being produced inside an archegonium? _____ If you focus up and down very carefully on an archegonium, you may be able to see an egg inside.

5. Examine a whole mount slide of a fern prothallium-antheridia. What is being produced inside the antheridia? _____ When sperm produced by the antheridia swim to the archegonia in a film of water, what results? _____ The latter develop into what generation? _____

The Life Cycle of a Fern

1. Which generation in the fern is dependent for any length of time on the other generation? _____

_____

2. Which generation is dispersed in ferns? _____ How? _____

Adaptation of Ferns to a Land Environment

1. State a significant way in which the fern is adapted to life on land, when you compare it to the moss. _____

2. List one characteristic of the fern illustrating that sexual reproduction is not adapted to a land environment. _____

14.3   Seed Plants

Seed plants (gymnosperms and angiosperms) are further adapted to live and reproduce on land. The dominant sporophyte contains vascular tissue, which not only transports water but also serves as an internal skeleton, allowing these plants to oppose the force of gravity. For example, all of today's trees are seed plants.

Seed plants are no longer dependent on external water to ensure fertilization. To understand the mechanism by which this has come about, it is necessary to know that seed plants have two types of sporangia, termed **microsporangia** and **megasporangia** (Fig. 14.9). In microsporangia, microspore mother cells produce microspores by meiosis and each develops into a male gametophyte generation, the **pollen grain.** During **pollination,** pollen grains are dispersed to the vicinity of female gametophytes.

In a megasporangium inside an **ovule,** a megaspore mother cell produces megaspores by meiosis. Only one of these develops into a female gametophyte that produces an egg. Following fertilization, the 2n zygote is still within the ovule, which becomes a seed. The seed is dispersed to a new location.

Seed Plants

1. Seed plants have two dispersal events. These events are dispersal of _____ and dispersal of _____. In gymnosperms, pollen grains are dispersed by the wind; in angiosperms, dispersal is sometimes by the wind and sometimes by insects or other animals.

2. During dispersal of seeds, what generation is dispersed to a new location? _____

Figure 14.9   Dispersal in seed plants.
(*top*) Pollen grains are produced in microsporangia and are dispersed during pollination. After reaching the female gametophyte, a pollen grain germinates and has a pollen tube. A sperm travels down the pollen tube to the egg. (*bottom*) An ovule, at first, contains a megasporangium, then a megaspore, which develops into a female gametophyte that produces an egg. Following fertilization, the ovule becomes the seed, which is dispersed to a new location.

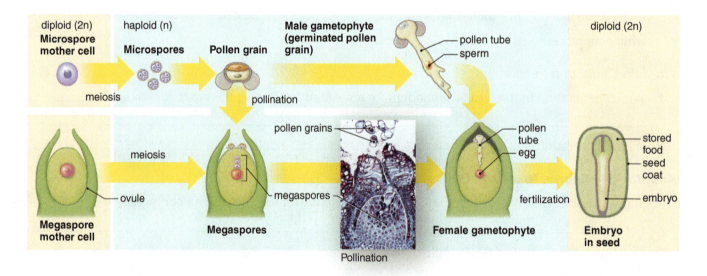

Gymnosperms

The gymnosperms are usually evergreen trees in which the sporangia are found on **cones** (Fig. 14.10 and Fig. 14.11). The **conifers** are by far the largest group of gymnosperms. Pines, spruces, firs, cedars, hemlocks, redwoods, and cypresses are all conifers. The pine has been chosen as representative of these plants.

Figure 14.10   Life cycle of pine.

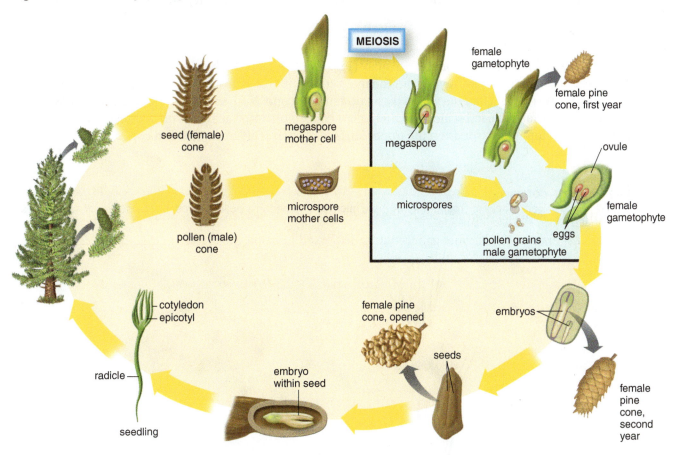

Figure 14.11   Conifers.
a. Vast areas of northern temperate regions are covered in evergreen coniferous forests. The tough, needlelike leaves of pines conserve water because they have a thick cuticle and recessed stomata. **b.** Conifers bear two types of cones: pollen (male) cones and seed (female) cones. **c.** The seed cones of junipers are fleshy.

a. A northern coniferous forest

b. Cones of lodgepole pine, *Pinus contorta*

c. Juniper, *Juniperus,* seed cones are fleshy.

Life Cycle of Pine Trees

1. Describe the sporophyte in the life cycle of the pine tree. _____
2. The sporophyte produces pollen (male) cones and seed (female) cones. Two microsporangia are located on the underside of each scale from the pollen cone. Here, microspore mother cells produce microspores that develop into pollen grains, the male gametophyte generation. Two ovules (megasporangia) are located on the underside of each scale from the seed cone. Here, megaspore mother cells produce four cells, one of which survives and becomes a megaspore.

Observation: Pine Cones

1. Observe the pollen and seed cones on display (Fig. 14.12). Compare their relative size and structure. Remove a single scale (sporophyll) from the male cone and from the seed cone, which has been heated so that the cone opens. Observe with a dissecting microscope. Note the two microsporangia located on the lower surface of each scale from the pollen cone. Note also the two ovules (megasporangia) that may have developed into seeds on the upper surface of each scale from the seed cone.

Figure 14.12   Pine cones.
a. The scales of pollen cones bear microsporangia, where microspores become pollen grains. **b.** The scales of seed cones bear ovules that develop into winged seeds.

a. Pollen cones

microsporangium

scale

b. Seed cone

seed

wing

scale

2. Examine the prepared slide of a longitudinal section through a mature pollen cone. On the lower surface of each scale are the microsporangia in which the microspore mother cells produce microspores. Each microspore develops into a male gametophyte, a pollen grain. Under high power, focus on a pollen grain and note the external wings and interior cells. One of these will divide to produce two cells, one of which is the sperm nucleus and the other of which forms the pollen tube, through which a sperm travels to the egg. *Label Figure 14.13.*

Figure 14.13 Pollen cone.
Pollen cones bear **a.** microsporangia in which microspores develop into pollen grains. **b.** Enlargement of pollen grains.

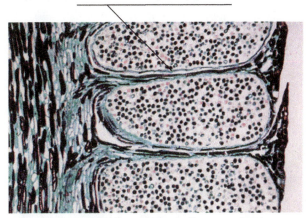

a. Longitudinal section through pine pollen cone, showing pollen grains within microsporangia

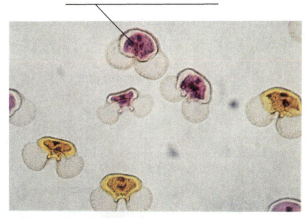

b. Enlargement of pollen grains

3. Examine the prepared slide of a longitudinal section through a seed cone and *label Figure 14.14.* In some ovules, you will see a megaspore mother cell surrounded by nutritive cells. You may also be able to find pollen grains just outside or within the integuments of an ovule. At about the time the megaspore mother cell is undergoing meiosis, the scales swell and open to allow the wind-dispersed pollen grains to enter. The megaspore undergoes a series of mitotic divisions and develops into the female gametophyte that contains two archegonia, each of which encloses a single large egg. What

generation is now within the ovule? _____

Figure 14.14 Seed cone.
Seed cones bear ovules, each of which at one time contains a megasporangium, shown here in longitudinal section. Note pollen grains near the entrance.

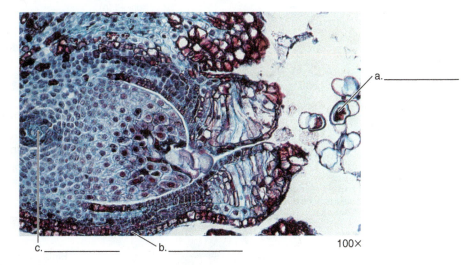

a. _____

100×

c. _____   b. _____

4. Following fertilization, a seed contains the embryonic plant and nutrient material within a seed coat. If available, examine pine seeds. The seeds of gymnosperms are windblown. In seed plants, seeds disperse the sporophyte. The word *gymnosperm* means "naked seeds," and the seeds themselves are not covered. If you wish, dissect a seed and, with the help of a hand lens, attempt to find the embryo.

Angiosperms

The angiosperms are the flowering plants. The flower contains the microsporangia and megasporangia. The word *angiosperm* means covered seeds, and these plants have seeds enclosed in fruits. The success of angiosperms is exemplified by their large number compared to that of the other plants.

Life Cycle of Flowering Plants

Study the diagram of a flowering plant life cycle (Fig. 14.15) and find the sporophyte generation. Describe the sporophyte, if the flowering plant is a deciduous tree (loses leaves in fall). _____

_____

Describe the sporophyte of a garden plant. _____

Figure 14.15 Life cycle of flowering plants.

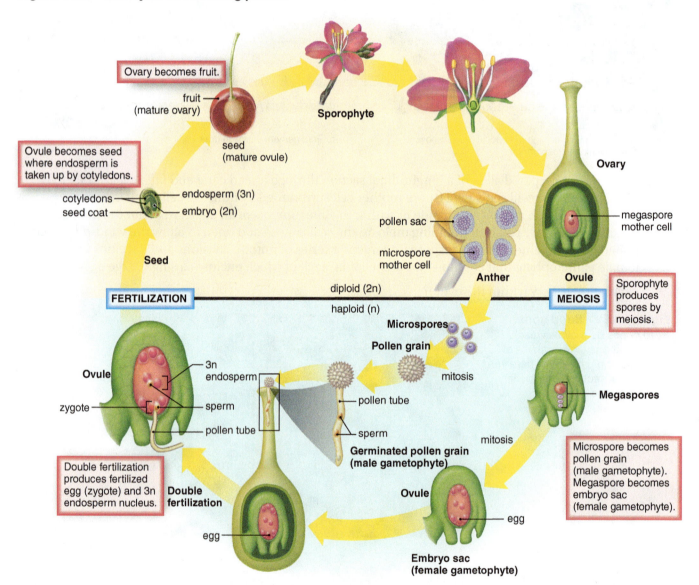

Figure 14.15 shows that each mother cell within the anther undergoes meiotic cell division to produce four haploid cells, the microspores. Each microspore divides mitotically to give a male gametophyte, a pollen grain. Following **pollination,** the transfer of pollen from the anther to the stigma, a pollen grain germinates and produces a long **pollen tube** in which two sperm travel to reach the ovule.

Inside each ovule, a megaspore mother cell undergoes meiosis to produce four megaspores, three of which disintegrate. The one remaining undergoes mitosis to give a multicelled female gametophyte. Flowering plants practice **double fertilization** because the other sperm joins two polar nuclei to form endosperm (3n), which serves as food for the developing embryo. Now the ovule becomes a seed. A seed contains the embryonic sporophyte and stored food within a seed coat. The seeds of angiosperms are covered by the ovary. The ovary becomes the fruit. All angiosperms produce fruit.

Observation: Flower

1. Identify these parts of a flower on a model using Figure 14.16.

   Sepal: the outermost set of modified leaves, collectively termed the **calyx.** The sepals are green in most flowers.

   Petal: the inner leaves that collectively comprise the **corolla.** The petals tend to be brightly colored.

   Stamen: comprised of a swollen terminal **anther** and a slender supporting filament.

   Carpel: comprised of a swollen basal **ovary;** a long, slender **style;** and a terminal **stigma.**

   Ovule: structure within ovary that contains the megasporangium and develops into the seed.

Figure 14.16   Anatomy of a typical flower.

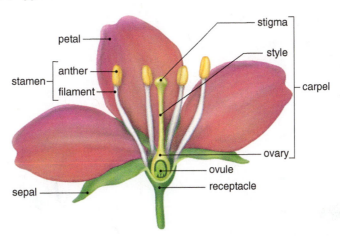

2. Carefully inspect a living flower. Remove the sepals and petals by breaking them off at the base. Are the stamens taller than the carpel? _____ Would self-fertilization be possible in this flower? _____

3. Remove a stamen and touch the anther to a drop of water on a slide. If nothing comes off in the water, crush the anther a little to squeeze out some of its contents. Place a coverslip on the drop and observe with low and high powers of the microscope. The somewhat spherical cells with thick walls are pollen

   grains. How is pollen dispersed? _____

   Flowering plants provide nectar to insects, whose mouth parts are adapted to acquiring the nectar from this particular species of plant. This is a mutualistic relationship. What does a pollinator do for

   the plant? _____

4. Remove the carpel by cutting it free just below the base. Make a series of thin cross sections through the ovary. The ovary is hollow and in this cavity there are small, nearly spherical bodies much larger than pollen grains. These are ovules. Remove an ovule that may be still attached by a stalk to the ovary wall. Place the ovule on a drop of water on a slide; cover with a cover glass; press firmly on the top of the cover glass with a clean eraser so as to smear the ovule into a thin mass.

   Observe with the microscope. Do you see cells within the ovule? _____

5. At maturity, an angiosperm seed contains the embryo and possibly some nutrient material in the form of endosperm covered by a seed coat. A fruit is a ripened ovary, together with any accessory flower parts that may be associated with the ovary. If available, examine a pea pod and an apple. Find the remnants of the flower parts still attached to the fruit. Open the pea pod and slice the apple. Which portion of each should be associated with the ovules; which with the ovary? _____

   _____

   Label flower remnants, fruit, and seed in the following diagram. Can you think of a biological advantage

   to producing fruits? _____ To producing a fleshy fruit?_____

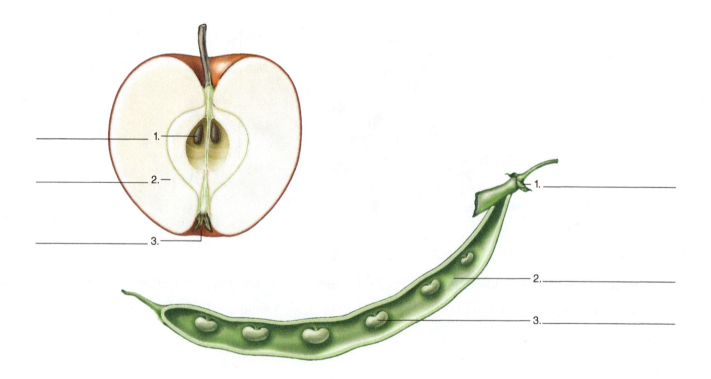

Compare the Life Cycle of a Pine Tree to That of a Flowering Plant

Complete the following table:

	Dominant Generation	Vascular Tissue (Present or Absent)	Dispersal of Sporophyte	Fruit (Present or Absent)
Conifer				
Flowering plant				

Adaptation of Seed Plants to a Land Environment

1. Do seed plants have vascular tissue that transports water and nutrients to all parts of the plant? _____

2. Can seed plants support a large body against the pull of gravity? _____ Explain.
   _____

3. Do seed plants reproduce without dependence on external water? _____ Explain.
   _____

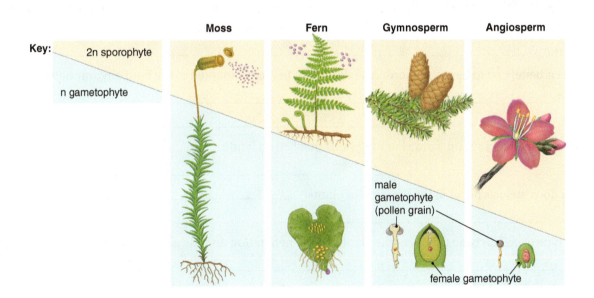

Comparison of Moss, Fern, Pine, and Flowering Plants

1. The preceding diagram tells you that the size of the _____ became progressively _____ as plants became adapted to live on land. Why is it suitable for this generation to be dependent on the sporophyte? _____

2. Which plants disperse spores? _____

3. Which plants have male and female gametophytes and disperse seeds? _____
   _____

4. Why is it appropriate to refer to "bees and plants" to explain sexual reproduction? _____
   _____

Laboratory 14 Plant Evolution   **179**

1. Name one way that each type of plant shows an adaptation to the land environment not seen before:

    a. Moss _____

    b. Lycophyte _____

    c. Fern _____

    d. Pine tree _____

    e. Flowering plant _____

2. The gametophyte is not protected in which of the plants listed in question 1?

    _____

3. In what way do flowering plants protect the female gametophyte? _____

    _____

    The male gametophyte? _____

4. In the flowering plant life cycle

    a. What becomes of the ovule? _____

    b. What becomes of the ovary? _____

5. Windblown spores are an adaptation to the land environment in which three types of plants studied in this

    laboratory? _____

6. Why is it beneficial to have the sporophyte dominant in plants adapted to the land environment?

    _____

7. Which generation is in a seed? _____

8. Why are flagellated sperm a drawback in plants that live on land? _____

    _____

9. When does meiosis occur in the life cycle of plants? _____

    _____

10. What type of cell division produces the gametophyte generation and the gametes in plants? _____

    _____

15

Plant Anatomy and Growth

Learning Outcomes

15.1 Plant Organs
- Distinguish between the root system and the shoot system of a plant.
- Identify the external anatomical features of a flowering plant.
- State the functions of a leaf, a stem, and a root.
- List five differences between monocots and eudicots.

15.2 Organization of Roots
- Name the zones of a eudicot root tip.
- Identify a cross section and the specific tissues of a eudicot root.

15.3 Xylem Transport
- Explain the continuous water column in xylem.

15.4 Organization of Stems
- Identify a cross section and the specific tissues of a herbaceous eudicot and herbaceous monocot stems and a woody stem.
- Distinguish between primary and secondary growth of a stem.
- Explain the occurrence of annual rings, and determine the age of a tree from a trunk cross section.

15.5 Organization of Leaves
- Distinguish between a monocot and a eudicot leaf.
- Identify a cross section and the specific tissues of a eudicot leaf.

Introduction

Despite their great diversity in size and shape, all flowering plants have three vegetative organs that have nothing to do with reproduction: the leaf, the stem, and the root. Leaves carry on photosynthesis and thereby produce the nutrients that sustain a plant. A stem usually supports the leaves so that they are exposed to sunlight. Roots anchor a plant and absorb water and minerals from the soil.

Each of these organs contains various tissues, arranged differently depending on whether a flowering plant is a monocot or a eudicot. The arrangement of tissues is distinctive enough that you should be able to identify the plant as a monocot or eudicot when examining a slide of a leaf, stem, or root.

Another way of grouping plants is according to whether they are herbaceous or woody. All flowering trees are woody; their stems contain wood. Many flowering garden plants and all grasses are herbaceous (nonwoody). Herbaceous plants have only **primary growth,** which increases their height. Woody plants have both primary and secondary growth. **Secondary growth** increases the girth of a tree. Only eudicots are ever woody.

Xylem is the vascular tissue that transports water from the roots to the leaves. The cohesion-tension model of xylem transport explains how the continuous column of water in xylem is able to rise to the top of a tall tree. During transpiration, water evaporates through openings in leaves called stomata (sing., stoma). This creates a tension that pulls the water column up because of the cohesive property of water.

15.1 Plant Organs

Figure 15.1 shows that a plant has a root system and a shoot system. The **shoot system** consists of the stem and leaves. The **root system** consists of a primary root and all of its lateral (side) roots.

Observation: A Living Plant

Shoot System

What is the primary function of the shoot system? _____

_____

The Leaves

1. Describe the **blade.**

_____

2. Describe the **petiole.**

_____

The Stem

1. Observe the **stem.** Locate a **node** and an **internode.**
2. Measure the length of the internode in the middle of the stem. Does the internode get larger or smaller toward the apex of the stem?

_____

Toward the roots? _____
Based on the fact that a stem elongates as it grows, explain

your observation. _____

_____

Figure 15.1 The body of a plant.
A plant has a root system, which extends below ground, and a shoot system, composed of the stem and leaves.

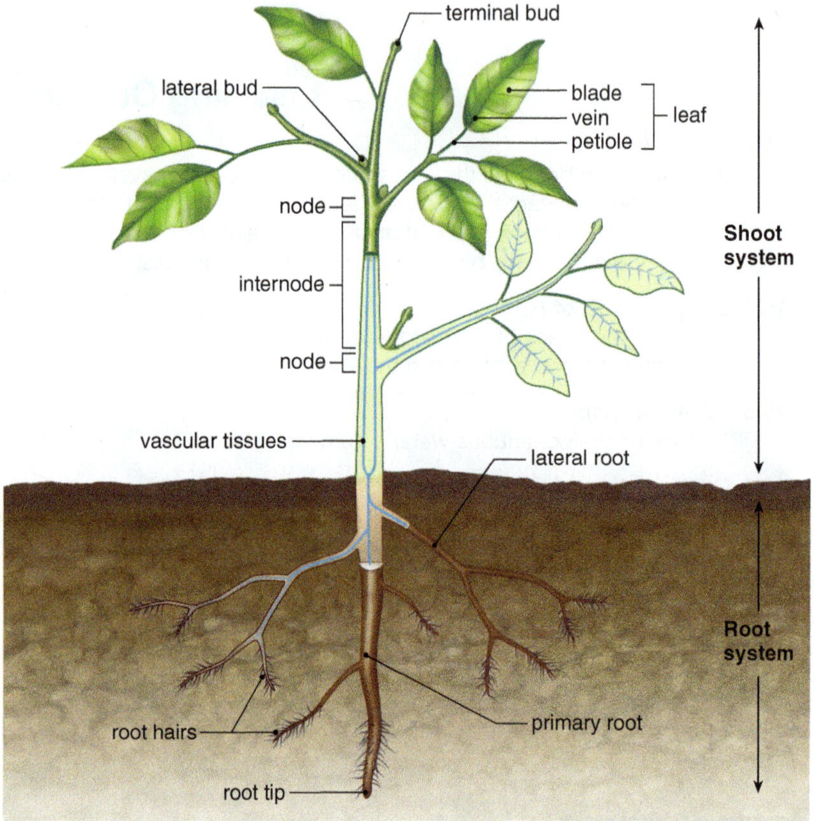

3. Where is the **terminal bud** of a stem? _____ Where is the **lateral (axillary) bud?** _____ Buds produce new growth because they contain a plant tissue called meristem. **Meristem** is a plant tisue consisting of undifferentiated cells that can continuously divide to produce cells that later become specialized.

Root System

Observe the root system of a living plant if the root system is exposed. Does the plant have a taproot system—that is, a main root many times larger than the lateral roots? _____ Or does the plant have a fibrous root system—that is, all the roots approximately the same size? _____ What is the primary function of the root system? _____

Figure 15.2   Flowering plants are either monocots or eudicots.
Five features are typically used to distinguish monocots from eudicots: number of cotyledons; arrangement of vascular tissue in roots, stems, and leaves; and number of flower parts.

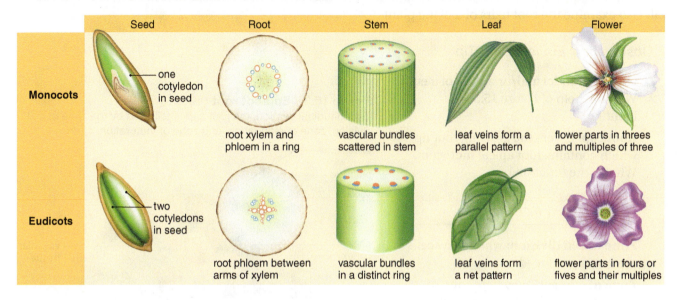

Monocots Versus Eudicots

Flowering plants are classified into two groups: **monocots** and **eudicots.** These terms refer to the number of cotyledons in the seed. A cotyledon is a seed leaf that contains nourishment for the growing embryo. Monocots have one cotyledon and eudicots have two cotyledons. In this laboratory, you will be studying the differences between monocots and eudicots with regard to the leaves, stems, and roots, as noted in Figure 15.2.

Observation: Monocot Versus Eudicot

1. Observe the leaves of the plant you are studying. Based on Figure 15.2, is this plant a monocot or a eudicot? _____ Explain. _____

2. Observe any other available types of leaves, and note in Table 15.1 the name of the plant and whether it is a monocot or a eudicot.

Table 15.1 Monocots Versus Eudicots		
Name of Plant	**Organization of Leaf Veins**	**Monocot or Eudicot?**
1		
2		
3		
4		

15.2 Organization of Roots

First you will study a eudicot root tip in longitudinal section and then a eudicot root in cross section. You will also examine the root hairs of a living plant.

Observation: Eudicot Root Tip

1. Obtain a model and/or a slide of a eudicot root tip.
2. With the help of Figure 15.3, identify

   Root cap: covers the growing root tip, which contains root apical meristem. What is the function of

   the root cap? _____

   _____

   Zone of cell division: where new cells are being produced.

   Zone of elongation: where rows of newly produced cells elongate. Which two zones are responsible for growth of the root tip?

   _____

   _____

   Zone of maturation: where the cells are specialized to carry on a particular function. You can recognize the zone of maturation because of the presence of root hairs. **Root hairs** increase the area for absorption of what by a root?

   _____

Figure 15.3 Eudicot root tip.
In longitudinal section, the root cap is followed by the zone of cell division, zone of elongation, and zone of maturation.

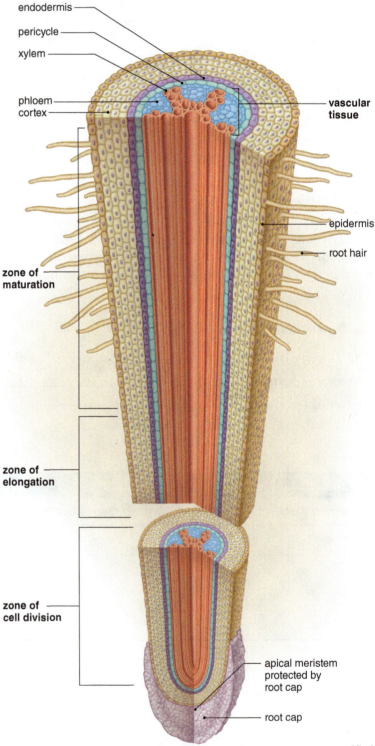

endodermis
pericycle
xylem
phloem
cortex
vascular tissue
epidermis
root hair
zone of maturation
zone of elongation
zone of cell division
apical meristem protected by root cap
root cap

Figure 15.4 Eudicot root cross section.

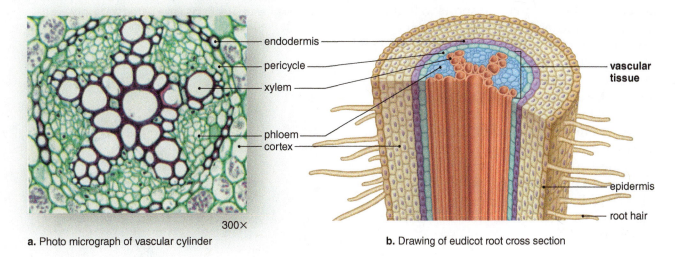

a. Photo micrograph of vascular cylinder

endodermis
pericycle
xylem
phloem
cortex

300×

vascular tissue

epidermis
root hair

b. Drawing of eudicot root cross section

Observation: Eudicot Root Slide

1. Obtain a slide of a eudicot root, such as *Ranunculus* (buttercup) root, in cross section.
2. Identify
   Epidermis: the outermost layer of small cells that gives rise to **root hairs.** Notice in Figure 15.4 that the epidermis is a single layer of cells.
   Cortex: just inside the epidermis. How many layers of thin-walled cells are present? _____
   The cortex stores the products of photosynthesis. *Label starch grains in Figure 15.4a.*
   Endodermis: a single layer of cells following the cortex. Endodermal cells have a waxy layer, called the **Casparian strip,** on four sides. This means that water and minerals have to pass through endodermal cells and not around them. For this reason, the endodermis regulates what materials enter a plant through the root.
   Pericycle: a layer one or two cells thick, just inside the endodermis. Lateral roots originate from this tissue.
   Vascular tissue (xylem and phloem): **Xylem** has several "arms" that extend like the spokes of a wheel. This tissue conducts water and minerals from the roots to the stem. **Phloem** is located between the arms of xylem. Phloem conducts organic nutrients from the leaves to the roots and other parts of the plant.
3. With the help of Figure 15.4, trace the path of water as it crosses a root from a root hair to
   xylem: _____
   _____

Observation: Root Hairs

1. Obtain a young germinated seedling, and float it in some water in a petri dish while you observe it.
2. Use the binocular dissecting microscope, and locate the root tip. Note the root cap and the region where the root hairs have formed on the root surface. What proportion of the root has
   root hairs? _____
3. Remove the root from the seedling, and make a wet mount of the root, using 0.1% neutral red.
4. Observe your slide under the microscope. Does every epidermal cell have a root hair? _____
   How do root hairs aid absorption? _____

15.3 Xylem Transport

Xylem (Fig. 15.5), which transports water from the roots to the leaves, contains two types of conducting cells: **tracheids** and **vessel elements.** Both types of conducting cells are hollow and nonliving, the vessel elements are larger, they lack transverse end walls, and they are arranged to form a continuous pipeline for water and mineral transport. The water column in xylem is continuous because water molecules are cohesive (they cling together) and because water molecules adhere to the sides of xylem cells. Therefore, water evaporation from leaf surfaces creates a negative pressure that pulls water upward.

Figure 15.5   Xylem structure.
Xylem contains two types of conducting cells: tracheids and vessel elements. Tracheids have pitted walls, but vessel elements are larger and form a continuous pipeline from the roots to the leaves.

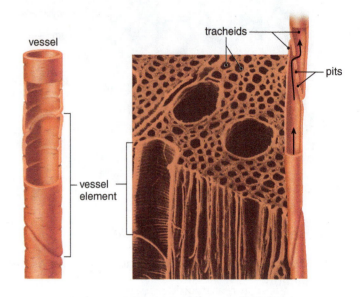

Experimental Procedure: Xylem Transport

1. Using an dropper, place a small amount of *red-colored water* in two beakers. Label one beaker "wet" and the other beaker "dry."
2. Transfer a stalk of *celery* (which was cut and then immediately placed in a container of water) into the "wet" beaker so that the large end is in the colored water.
3. Transfer a stalk of *celery* of approximately the same length and width (but that was kept in the air after being cut) into the "dry" beaker so that the large end is in the colored water.
4. With scissors, cut off the top end of each stalk, leaving about 10 centimeters total length.
5. Time how long it takes for the red-colored water to reach the top of each stalk, and record these data in Table 15.2.
6. In which celery stalk was the water column broken? _____

   _____
   Use this information to write a conclusion in Table 15.2.
7. Make a thin, cross-sectional wet mount of the stalk in the "wet" beaker. Observe this slide under the microscope. What type of tissue has been stained by the dye?

Table 15.2 Celery Stalk Experiment		
Stalk	**Speed of Dye (Minutes)**	**Conclusion**
Cut end placed in water prior to experiment		
Cut end kept in air prior to experiment		

15.4 Organization of Stems

Stems are usually found aboveground and provide support for leaves and flowers. Stems that do not contain wood are called **herbaceous,** or nonwoody, stems. Herbaceous stems undergo primary growth but they do not undergo secondary growth. Activity of meristem in a terminal bud results in primary growth of a stem.

Anatomy of Herbaceous Stems

Usually, monocots remain herbaceous throughout their lives. Some eudicots, such as those that live a single season, are also herbaceous. Other eudicots, such as trees, become woody as they mature.

Observation: Anatomy of Eudicot and Monocot Herbaceous Stems

1. Observe a prepared slide of a eudicot stem, such as *Helianthus* (sunflower). With the help of Figure 15.6*a*, identify

   Epidermis: the outermost protective layer.

   Cortex: may aid photosynthesis and store nutrients.

   Vascular bundles transport water and organic nutrients. In a herbaceous eudicot stem the vascular bundles occur in a ring pattern.

   Pith: stores organic nutrients.

2. Observe a prepared slide of a monocot stem, such as *Zea mays* (corn) (Fig. 15.6*b*). The vascular bundles are scattered. *Joint label the vascular bundles in Fig. 15.6a and b.*

3. State the main difference between the eudicot stem and the monocot stem. _____

_____

Figure 15.6 Nonwoody (herbaceous) stems.
a. In eudicot stems, the vascular bundles are arranged in a ring around well-defined ground tissue called pith.
b. In monocot stems, the vascular bundles are scattered within the ground tissue.

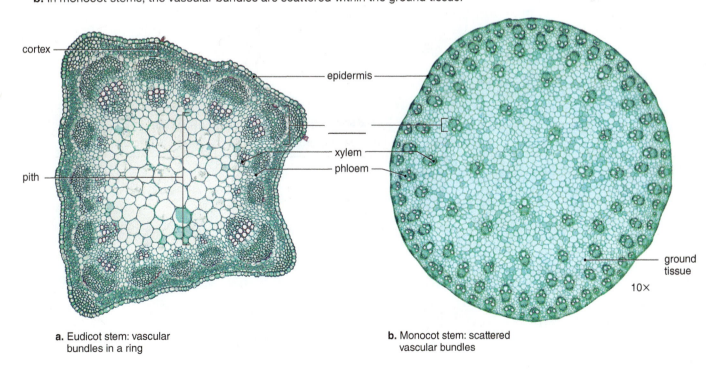

a. Eudicot stem: vascular bundles in a ring

b. Monocot stem: scattered vascular bundles

Laboratory 15 Plant Anatomy and Growth

Anatomy of Woody Stems

Woody stems undergo both primary growth (increase in length) and secondary growth (increase in girth). When *primary growth* occurs, **shoot apical meristem** within a terminal bud is active and **root apical meristem** within a root tip is active. When *secondary growth* occurs, vascular cambium is active. **Vascular cambium** is meristem tissue, which produces new xylem and phloem called **secondary xylem** and **phloem** each year. The buildup of secondary xylem year after year is called **wood.**

Observation: Anatomy of Winter Twig

1. A winter twig typically shows several years' past primary growth. Examine two examples of winter twigs (Fig. 15.7), and identify the **terminal bud** located at the tip of the twig. Primary growth of a stem originates here because a terminal bud contains shoot apical meristem.
2. Locate a **terminal bud scar.** These scars encircle the twig and indicate where the terminal bud was located in previous years. The distance between two adjacent terminal bud scars equals one year's primary growth.
3. Find a **leaf scar.** This is where the petiole of a leaf was attached to the stem.
4. Identify a **node.** This is where a leaf scar is located.
5. Note the **vascular bundle scars** in the leaf scar. Complete this sentence: Vascular bundle scars appear where

   the vascular bundles _____.
6. Locate a **lateral bud.** This is where new branch growth can occur because a lateral bud also contains meristem tissue.

Figure 15.7 External structure of a winter twig.
Counting the terminal bud scars tells the age of a particular branch.

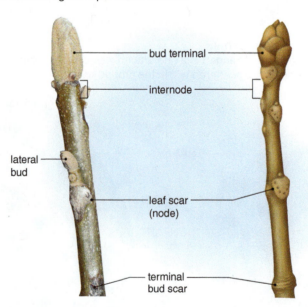

1. Examine a prepared slide of a cross section of a woody stem, such as *Tilia* (tulip tree) (Fig. 15.8), and identify the **bark** (the dark, outer area), which contains **cork,** a protective outer layer; **cortex,** which stores nutrients; and **phloem,** which transports organic nutrients.
2. Locate the **vascular cambium** at the inner edge of the bark, between the bark and the wood.
3. Find the **wood,** which contains annual rings. An **annual ring** is the amount of xylem added to the plant during one growing season. Rings appear to be present because spring wood has large xylem vessels and looks light in color, while summer wood has much smaller vessels and appears much darker. How old is the stem you are observing? —————— Are all the rings the same width? —————
4. Identify the **pith,** a ground tissue that stores organic nutrients and may disappear.
5. Locate **rays,** groups of small, almost cuboidal cells that laterally extend out from the pith.

Figure 15.8 Woody eudicot stem cross section.
Because xylem builds up year after year, it is possible to count the annual rings to determine the age of a tree. This tree is three years old.

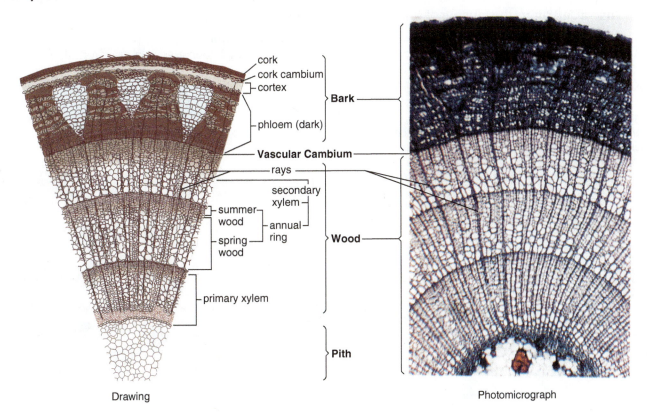

Drawing

Photomicrograph

15.5 Organization of Leaves

Leaves are generally broad and quite thin to better capture solar energy. Carbon dioxide enters a leaf at openings called **stomata** (sing., stoma), and water enters by way of **leaf veins,** extensions of the vascular bundles from the stem.

Observation: Stomata

1. Obtain a *leaf* from a plant designated by your instructor, and put a drop of *distilled water* on a slide.
2. Using the technique shown in Figure 15.9a, obtain a strip of outer tissue from a leaf. This tissue is epidermis.
3. Put the tissue in the drop of water on the slide, outer side up. Add a coverslip, and examine it microscopically, using both low power and high power.
4. Observe a stoma and the two guard cells that regulate the opening and closing of the stoma (Figure 15.9b).

    What gas enters a leaf at the stomata? _____

5. Count the number of stomata in the high-power field of view: _____.
6. Assume that the area of the high-power field is 0.10 mm^2. Divide the number of stomata by this

    area to determine the number of stomata in 1 square millimeter: _____.

7. Does your leaf contain a large number of stomata per square millimeter? _____

Figure 15.9 Stomata.
a. Method of obtaining a strip of epidermis from the underside of a leaf. **b.** Photo to help you identify stomata under the microscope. Each stoma is bordered by two guard cells.

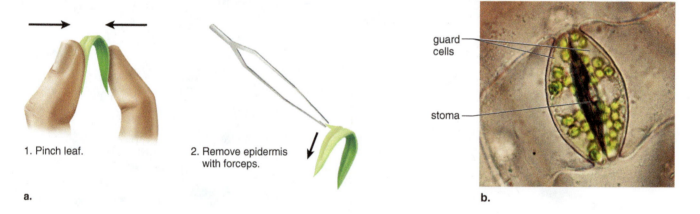

1. Pinch leaf.

2. Remove epidermis with forceps.

guard cells

stoma

a.

b.

1. Obtain a slide of a leaf, such as *Liqustrum* (common privet) or *Syringa* (barley), in cross section.
2. With the help of Figure 15.10, identify

Cuticle: the outermost layer that protects the leaf and prevents water loss.

Upper and lower epidermis: a single layer of protective cells at the upper and lower surfaces.

Leaf vein: transports water and organic nutrients. What tissues does a leaf vein contain?

_____ and _____

Palisade mesophyll: located near the upper epidermis. *Label the palisade mesophyll in Figure 15.10.*

Spongy mesophyll: located near the lower epidermis. *Label the spongy mesophyll in Figure 15.10.*

Which type of mesophyll has chloroplasts? _____

Which type of mesophyll carries on photosynthesis? _____

Which type of mesophyll has air spaces that facilitate exchange of gases? _____

Stomata: the openings you examined in the previous observation.

Figure 15.10 Leaf anatomy.
Complete the labels as directed in the Observation on this page.

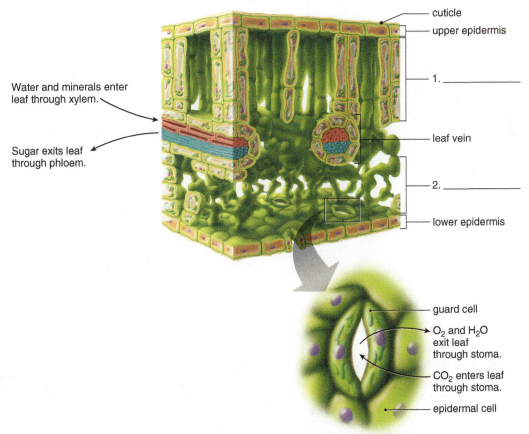

Stoma and guard cells

1. Given the information in this laboratory, how would you distinguish between a monocot plant and a eudicot plant based on their external anatomy? _____ _____

2. What is meristem tissue? _____ How is this tissue different from all other types of plant tissue? _____

3. In which zone of a eudicot root would you expect to find vascular tissue? Why? _____ _____ _____

4. In a eudicot root, what structural feature allows the endodermis to regulate the entrance of water and materials into the vascular cylinder, where xylem and phloem are located? _____ _____

5. Characterize the root of a carrot. _____

6. How would you microscopically distinguish a eudicot stem from a monocot stem? _____ _____

7. Distinguish between primary and secondary growth of a woody stem, and explain how each arises. _____ _____

8. Contrast how you could determine one year's growth by looking at a winter twig with how you determine one year's growth in a cross section of a tree stem. _____ _____ _____

9. Contrast the manner in which water reaches the inside of a leaf with the manner in which carbon dioxide reaches the inside of a leaf. _____ _____

10. How would you recognize the epidermis of a root versus the epidermis of a leaf? _____ _____

Essentials of Biology Website

Instructors can find lab prep information and answers to all of the laboratory questions in the Laboratory Resource Guide. Students can practice their knowledge with quizzes, animations, flashcards, and much more.

www.mhhe.com/maderessentials4

McGraw-Hill Access Science Website

An online encyclopedia of science and technology that provides information, including videos, that can enhance the laboratory experience.

www.accessscience.com

16

Animal Evolution

Introduction

This laboratory concerns the animal kingdom. Animals are all multicellular and have varying degrees of motility. They are heterotrophic and most digest their food in a digestive cavity.

One of the themes of today's laboratory will be the similarities and differences between the animals studied. The molluscs (e.g., clam) and arthropods (e.g., crayfish and grasshopper) are invertebrates, while the amphibians (e.g., frog) are vertebrates. Invertebrates lack a backbone of vertebrae, which is present in vertebrates. All of the animals in today's laboratory have true tissues, bilateral symmetry, a complete digestive tract, and a coelom or body cavity. They differ in that molluscs are nonsegmented, while the arthropods and vertebrates are segmented animals with jointed appendages. Jointed appendages are particularly adapted for locomotion on land. The insects are arthropods that are plentiful on land. Among the vertebrates, the amphibians live on land at least part of their lives, while the reptiles, including birds, and mammals are primarily land animals (Fig. 16.1).

Figure 16.1 Animal diversity.
The chameleon, a reptile, and the fly, an arthropod, are both very well adapted to living on land. The skin of a reptile and the external skeleton of an arthropod resist drying out. They both breathe by taking in air and both have a suitable means of locomotion on land.

Figure 16.2   A modern look at the animal evolutionary tree.
The relationships in this diagram are based on research into the developmental biology of each group, as well as molecular studies of DNA, RNA, and protein similarity.

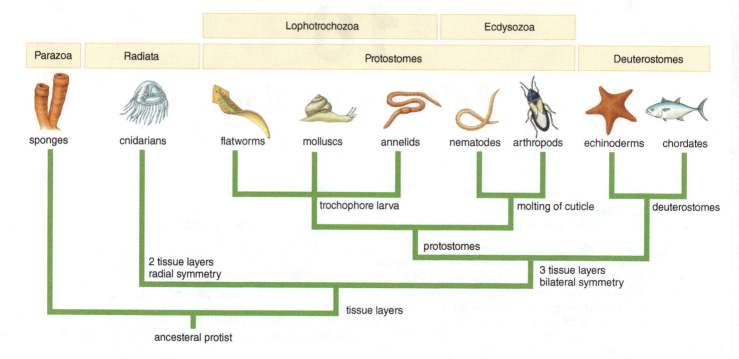

16.1   Evolution of Animals

Today, molecular data are used in addition to comparative anatomy to trace the evolutionary history of animals. These data tell us that all animals share a common ancestor. This common ancestor was most likely a choanoflagellate, a protist consisting of a colony of flagellated cells. All but one of the phyla depicted in the tree (Fig. 16.2) consist of only invertebrates—the chordates contain a few invertebrates as well as the vertebrates.

Certain anatomical features of animals are used in the tree. The first feature of interest is formation of tissue layers. Sponges have no true tissue layers. Which phyla in the tree have only two tissue layers?

_____ The other phyla have three tissue layers.

Another feature of interest is **symmetry. Asymmetry** means the animal has no particular symmetry. Which phyla have **radial symmetry,** in which, as in a wheel, two identical halves are obtained no matter

how the animal is longitudinally sliced? _____ The other

phyla have **bilateral symmetry,** which means the adult animal has a definite right half and left half.

Finally, complex animals are either **protostomes** (first opening during development is the mouth) or **deuterostomes** (second opening during development is the mouth [the first one is the anus]). Which pattern of development do the flatworms, rotifers, and nematodes (also called roundworms)

have? _____

16.2 Invertebrates

All of the phyla in Figure 16.2 contain invertebrates. The vertebrates only occur in **phylum chordata.** Most types of invertebrates are adapted to living in the sea, the insects being the major exception to this statement. We will have an opportunity to contrast adaptations to the land environment with adaptations to the aquatic environment.

Molluscs

Most **molluscs** are marine, but there are also some freshwater and terrestrial molluscs (Fig. 16.2). All molluscs have a three-part body consisting of a ventral, muscular **foot** specialized for various means of locomotion; a **visceral mass** that includes the internal organs; and a **mantle,** a thin tissue that encloses the visceral mass and may secrete a shell.

Observation: Diversity of Molluscs

Most molluscs belong to one of four groups. (1) The grazing marine herbivores, such as chitons, have a body flattened dorsoventrally covered by a shell consisting of eight plates (Fig. 16.3). (2) The bivalves contain marine and freshwater sessile filter feeders, such as clams, with a body enclosed by a shell consisting of two valves. These animals have a hatchet-shaped foot but no head or radula. (3) The cephalopods contain marine active predators, such as squids. The shell may be reduced or absent; the head, which is anterior to an elongated visceral mass, bears a circle of tentacles and arms. The circulatory system is efficient, and the well-developed nervous system is accompanied by cephalization. (4) The gastropods contain marine, freshwater, and often terrestrial herbivores, such as snails, with a coiled shell that distorts body symmetry. The head has tentacles.

1. Examine mollusc specimens, and complete the first two columns of Table 16.1.
2. Examine the foot, and complete the third column of Table 16.1. Some molluscs have a broad, flat foot, others a hatchet-shaped foot; in still others the foot has become tentacles and arms that assist in the capture of food.
3. Indicate in the last column if cephalization is present or not present.

Figure 16.3   Three groups of molluscs.
Top row: Snails (*left*) and nudibranchs (*right*) are gastropods. Middle row: Octopuses (*left*) and nautiluses (*right*) are cephalopods. Bottom row: Scallops (*left*) and mussels (*right*) are bivalves.

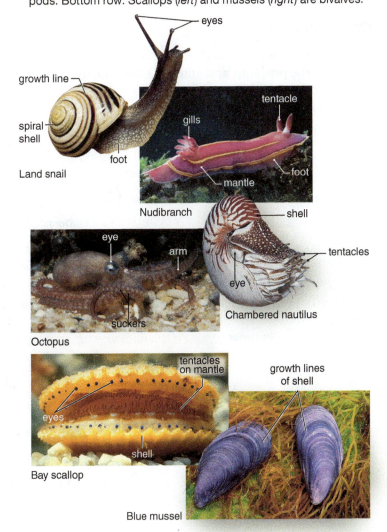

Table 16.1 Molluscan Diversity

Common Name of Specimen	Group	Description of Foot	Cephalization (Yes or No)

Anatomy of Clam

Clams are bivalved because they have right and left shells secreted by the mantle. Clams have no head, and they burrow in sand by extending a "hatchet" foot between the valves. Clams are filter feeders and feed on debris that enters the mantle cavity. Clams have an open circulatory system; the blood leaves the heart and enters sinuses (cavities) by way of anterior and posterior aortas. There are many different types of clams. The one examined here is the freshwater clam *Venus*.

Observation: Anatomy of Clam

External Anatomy

1. Examine the external shell (Fig. 16.4) of a preserved clam (*Venus*). The shell is an **exoskeleton.**
2. Find the posterior and anterior ends. The more pointed end of the **valves** (the halves of the shell) is the posterior end.
3. Determine the clam's dorsal and ventral regions. The valves are hinged together dorsally.
4. What is the function of a heavy shell? _____

Figure 16.4 External view of clam shell.

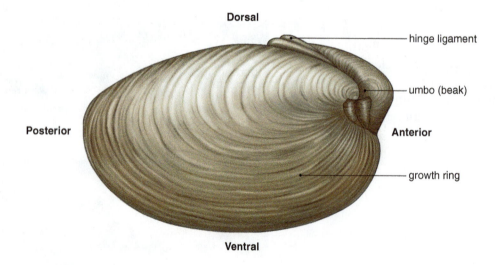

Internal Anatomy

1. Place the clam in the dissecting pan, with the **hinge ligament** and **umbo** (blunt dorsal protrusion) down. Carefully separate the **mantle** from the right valve by inserting a scalpel into the slight opening of the valves. What is a mantle? _____

_____

2. Insert the scalpel between the mantle and the valve you just loosened.
3. The **adductor muscles** hold the valves together. Cut the adductor muscles at the anterior and posterior ends by pressing the scalpel toward the dissecting pan. After these muscles are cut, the valve can be carefully lifted away. What is the advantage of powerful adductor muscles? _____

_____

4. Examine the inside of the valve you removed. Note the concentric lines of growth on the outside, the hinge teeth that interlock with the other valve, the adductor muscle scars, and the mantle line. The inner layer of the shell is mother-of-pearl.
5. Examine the rest of the clam (Fig. 16.5) attached to the other valve. Notice the mantle, which lies over the visceral mass and foot, and the adductor muscles.
6. Bring the two halves of the mantle together. Explain the term *mantle cavity*. _____

_____

7. Identify the **incurrent** (more ventral) and **excurrent siphons** at the posterior end (Fig. 16.5). Explain how water enters and exits the mantle cavity. _____

_____

8. Cut away the free-hanging portion of the mantle to expose the **gills.** Does the clam have a respiratory organ? _____ What type of respiratory organ? _____

9. A mucous layer on the gills entraps food particles brought into the mantle cavity, and the cilia on the gills convey these food particles to the mouth. Why is the clam called a filter feeder? _____

_____

10. The nervous system is composed of three pairs of ganglia (located anteriorly, posteriorly, and in the foot), all connected by nerves. The clam does not have a brain. A ganglion contains a limited number of neurons, whereas a brain is a large collection of neurons in a definite head region.
11. Identify the **foot,** a tough, muscular organ for locomotion, and the **visceral mass,** which lies above the foot and is soft and plump. The visceral mass contains the digestive and reproductive organs.
12. Identify the **labial palps** that channel food into the open mouth.
13. Identify the **anus,** which discharges into the excurrent siphon.
14. Find the **intestine** by its dark contents. Trace the intestine forward until it passes into a sac, the clam's only evidence of a coelom.
15. Locate the **pericardial sac (pericardium)** that contains the heart. The intestine passes through the heart. The heart pumps blood into the aortas, which deliver it to blood sinuses in the tissues. A clam has an **open circulatory system.** Explain. _____

_____

16. Cut the visceral mass and the foot into exact left and right halves, and examine the cut surfaces. Identify the greenish-brown digestive glands; the stomach, embedded in the digestive glands; and the intestine, which winds about in the visceral mass. Reproductive organs are also present.

Figure 16.5 Anatomy of a bivalve.
The mantle has been removed to reveal the internal organs. **a.** Drawing. **b.** Dissected specimen.

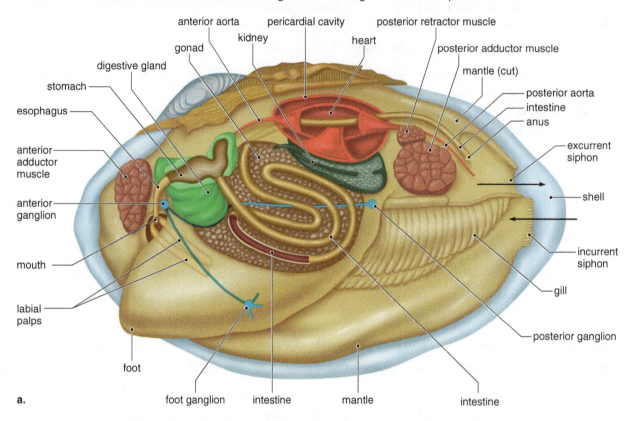

a.

b.

Anatomy of Squid

Squids are cephalopods (means head-foot) because they have a well-defined head. The head contains a brain and bears sense organs. The squid moves quickly by jet propulsion of water, which enters the mantle cavity by way of a space that circles the head. When the cavity is closed off, water exits by means of a funnel. Then the squid moves rapidly in the opposite direction.

The squid seizes fish with its tentacles; the mouth has a pair of powerful, beaklike jaws and a **radula,** a filelike organ containing rows of teeth. The squid has a closed circulatory system composed of vessels and three hearts, one of which pumps blood to all the internal organs, while the other two pump blood to the gills located in the mantle cavity.

Observation: Anatomy of Squid

1. Examine a preserved squid.
2. Refer to Figure 16.6 for help in identifying the mouth (defined by beaklike jaws and containing a radula) and the tentacles and arms, which encircle the mouth.
3. Locate the head with its sense organs, notably the large, well-developed eye.
4. Find the funnel, where water exits from the mantle cavity, causing the squid to move backward.
5. If the squid has been dissected, note the heart, gills, and blood vessels.

Clam Anatomy Compared with Squid Anatomy

1. Compare clam anatomy with squid anatomy by completing Table 16.2.
2. Explain how both clams and squids are adapted to their way of life.

_____

_____

_____

_____

_____

Table 16.2 Comparison of Clam and Squid		
	Clam	**Squid**
Feeding mode		
Skeleton		
Circulation		
Cephalization		
Locomotion		
Nervous system		

Figure 16.6 Anatomy of a squid.
The squid is an active predator and lacks the external shell of a clam. It captures fish with its tentacles and bites off pieces with its jaws. A strong contraction of the mantle forces water (arrows) out the funnel, resulting in "jet propulsion." **a.** Drawing. **b.** Dissected specimen.

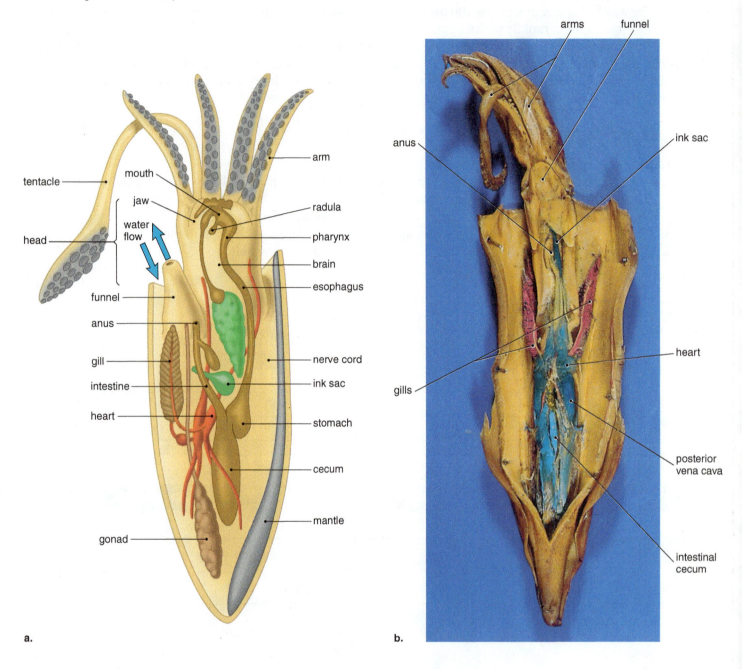

a.

b.

Arthropods

Arthropods have paired, jointed appendages and a hard exoskeleton that contains chitin. The chitinous exoskeleton consists of hardened plates separated by thin, membranous areas that allow movement of the body segments and appendages. Arthropods are segmented, but specialization of segments has occurred.

Explain. _____

_____

The arthropods are divided into three groups. In one group are the spiders, scorpions, and horseshoe crabs; in another are the millipedes, centipedes, and insects; and the third group consists of the crustaceans (e.g., crabs) and barnacles.

Observation: Diversity of Arthropods

Examine various specimens of arthropods (Fig. 16.7), and complete Table 16.3. In the last column, note the number of legs attached to the thorax.

Figure 16.7 Crustacean diversity.

a. A copepod uses its long antennae for floating and its feathery maxillae for filter feeding. **b.** Shrimp and **(c)** crabs are decapods—they have five pairs of walking legs. Shrimp resemble crayfish more closely than they do crabs, which have a reduced abdomen. Marine shrimp feed on codepods. **d.** Barnacles have no abdomen and a reduced head; the thoracic legs project through a shell to filter feed. Barnacles often live on humanmade objects, such as ships, buoys, and cables. The gooseneck barnacle attaches to an object by a long stalk.

Table 16.3 Arthropod Diversity		
Common Name of Specimen	**Group***	**Number of Legs (Attached to Thorax)**

*Choose crustacean (see Fig. 16.7), insects (see Fig. 16.10), arachnids (e.g., spiders, scorpions), centipedes, or millipedes.

Anatomy of Crayfish

Crayfish belong to a group of arthropods called crustaceans (Fig. 16.7). Crayfish are adapted to an aquatic existence. They are known to be scavengers, but they also prey on other invertebrates. The mouth is surrounded by appendages modified for feeding, and there is a well-developed digestive tract.

Observation: Anatomy of Crayfish

External Anatomy

1. Obtain a preserved crayfish, and place it in a dissecting pan.
2. Identify the chitinous **exoskeleton.** With the help of Figure 16.8, identify the head, thorax, and abdomen. Together, the head and thorax are called the **cephalothorax;** the cephalothorax is covered by the **carapace.** Has specialization of segments occurred? _____ Explain. _____

   _____

3. Find the **antennae,** which project from the head. At the base of each antenna, locate a small, raised nipple containing an opening for the **green glands,** the organs of excretion. Crayfish excrete a liquid nitrogenous waste.
4. Locate the **compound eyes,** composed of many individual units for sight. Do crayfish demonstrate

   cephalization? _____ Explain. _____

5. Identify the several pairs of appendages around the mouth for handling food. Find the five pairs of walking legs attached to the cephalothorax. The most anterior pair is modified as pincerlike claws.
6. Locate the five pairs of **swimmerets** on the abdomen. In males, the anterior two pairs are stiffened and folded forward. They are claspers that aid in the transfer of sperm during mating.
7. In females, identify the **seminal receptacles,** a swelling located between the bases of the third and fourth pairs of walking legs. Sperm from the male are deposited in the seminal receptacles. In the male, identify the opening of the sperm duct located at the base of the fifth walking leg.

   What sex is your specimen? _____ Examine the opposite sex also.

8. Find the last abdominal segment, which bears a pair of broad, fan-shaped **uropods** that, together with a terminal extension of the body, form a tail. Has specialization of appendages occurred in a

   crayfish? _____ Explain. _____

Figure 16.8   External anatomy of a crayfish.

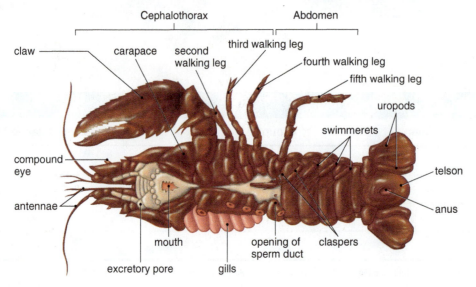

Figure 16.9   Internal anatomy of a crayfish.

ovary

stomach

compound eye

abdomen

antenna

claw

uropod
telson

gills

walking
leg

digestive
gland

Internal Anatomy

Observe a longitudinal section of a grasshopper if available on demonstration. Try to locate the structures shown in Figure 16.9. Can you find the ventral solid nerve cord?

Anatomy of Grasshopper

The grasshopper is an **insect** (Fig. 16.10). Insects are adapted to life on land. In insects with wings, such as the grasshopper, wings are attached to the thorax. Respiration is by a highly branched internal system of tubes, called tracheae.

Figure 16.10   Insect diversity.

Louse

Dragonfly

Wasp

Leafhopper

Butterfly

Beetle

External Anatomy

1. Obtain a preserved grasshopper (*Romalea*), and study its external anatomy with the help of Figure 16.11. Identify the head, thorax, and abdomen.
2. The grasshopper's **thorax** consists of three fused segments: the large anterior **prothorax,** the middle **mesothorax,** and the hind **metathorax.** Identify the first pair of legs attached to the prothorax. Then find the second pair of legs and the outer pair of straight, leathery **forewings** attached to the mesothorax. Finally, locate the third pair of legs and the inner, membranous **hind wings** attached to the metathorax. Each leg consists of five segments. The hind leg is well developed and used for jumping. How many pairs of legs are there? _____
3. Is locomotion in the grasshopper adapted to land? _____

   Explain. _____

   _____

   _____

4. Use a hand lens or dissecting microscope to examine the grasshopper's special sense organs of the head. Identify the **antennae** (a pair of long, jointed feelers); the **compound eyes;** and the three, dotlike **simple eyes.**
5. Remove the **mouthparts** by grasping them with forceps and pulling them out. Arrange them in order on an index card, and compare them with Figure 16.11*b*. These mouthparts are used for chewing and are quite different from those of a piercing and sucking insect.
6. Identify the **tympana** (sing., **tympanum**), one on each side of the first abdominal segment (Fig. 16.11*a*). The grasshopper detects sound vibrations with these membranes.
7. Locate the **spiracles,** along the sides of the abdominal segments. These openings allow air to enter the **tracheae** air tubes, which constitute the respiratory system. Tracheae take oxygen directly to the leg and wing muscles. Why is this beneficial? _____

   _____

8. Find the **ovipositors** (Fig. 16.12*a*), four curved and pointed processes projecting from the hind end of the female. These are used to dig a hole in which fertilized eggs are laid. The male has **claspers** that are used during copulation (Fig. 16.12*b*) and fertilization of eggs is internal.

Internal Anatomy

Observe a longitudinal section of a grasshopper if available on demonstration. Try to locate the structures shown in Figure 16.13. What is the function of Malpighian tubules? _____

Conclusion of Arthropods

Compare the adaptations of a crayfish with those of a grasshopper by completing Table 16.4. Put a star beside each item that indicates an adaptation to life in the water (crayfish) and to life on land (grasshopper). How many did you identify? _____ Check with your instructor to see if you identified the maximum number of adaptations.

Figure 16.11 External anatomy of a female grasshopper, *Romalea*.
a. The legs and wings are attached to the thorax. **b.** The head has mouthparts of various types.

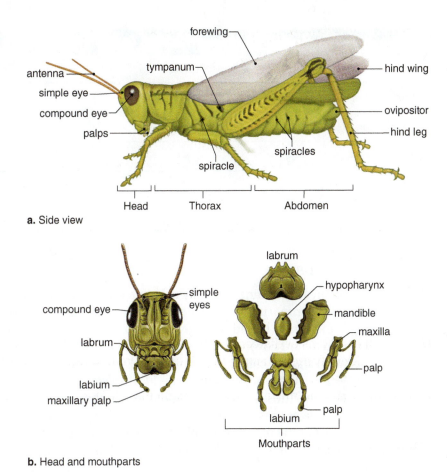

a. Side view

b. Head and mouthparts

Figure 16.12 Grasshopper genitalia.
a. Females have an ovipositor, and **(b)** males have claspers.

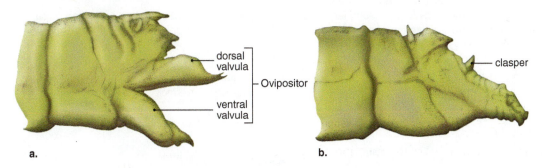

Figure 16.13 Internal anatomy of a female grasshopper.
The digestive system of a grasshopper shows specialization of parts.

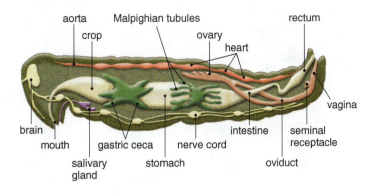

Table 16.4 Comparison of Crayfish and Grasshopper

	Crayfish	Grasshopper
Locomotion		
Respiration		
Nervous system		
Reproductive features		
Sense organs		

Insect Metamorphosis

Metamorphosis means a change, usually a drastic one, in form and shape. Some insects undergo what is called *complete metamorphosis,* in which case they have three stages of development: the larval stages, the pupa stage, and finally the adult stage. Metamorphosis occurs during the pupa stage, when the animal is enclosed within a hard covering. The animals best known for metamorphosis are the butterfly and the moth, whose larval stage is called a caterpillar and whose pupa stage is the cocoon; the adult is the butterfly or moth (Fig. 16.14*a*). Grasshoppers undergo *incomplete metamorphosis,* a gradual change in form rather than a drastic change. The immature stages of the grasshopper are called nymphs rather than larvae, and they are recognizable as grasshoppers even though they differ somewhat in shape and form (Fig. 16.14*b*).

If available, examine life-cycle displays or plastomounts that illustrate complete and incomplete metamorphosis. Other than those shown in Fig. 16.14, name a type of insect that undergoes complete

metamorphosis: _____and another that undergoes incomplete

metamorphosis: _____

Figure 16.14   Insect metamorphosis.
During **(a)** complete metamorphosis, a series of larvae leads to pupation. The adult hatches out of the pupa. During **(b)** incomplete metamorphosis, a series of nymphs leads to a full-grown grasshopper.

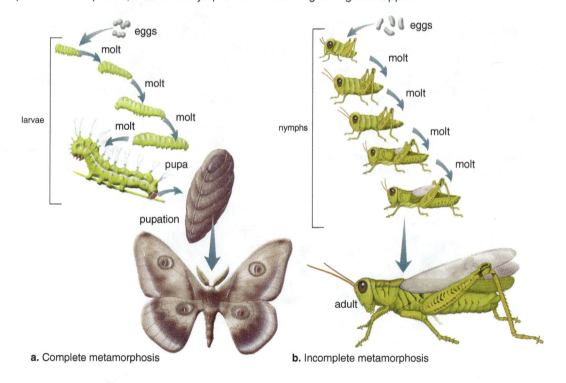

a. Complete metamorphosis   b. Incomplete metamorphosis

16.3 Vertebrates

All **chordates,** including vertebrates, have (1) a dorsal tubular nerve cord; (2) a dorsal supporting rod, called a notochord, at sometime in their life history; (3) a postanal tail (e.g., tailbone or coccyx); and, in chordates that breathe by means of gills, (4) pharyngeal pouches that become gill slits. In terrestrial chordates, these pouches are modified for other purposes.

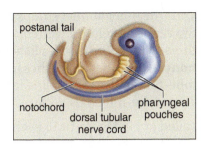

Evolutionary Tree

The evolutionary tree of chordates (Fig. 16.15) shows that some chordates, notably the **tunicates** and **lancelets,** are not vertebrates. These chordates retain a supporting rod called the **notochord** and are called the **invertebrate chordates.** Explain the term *invertebrate chordates.* _____

Figure 16.15 Evolutionary tree of the chordates.
Evolution of chordates is marked by these seven derived innovations.

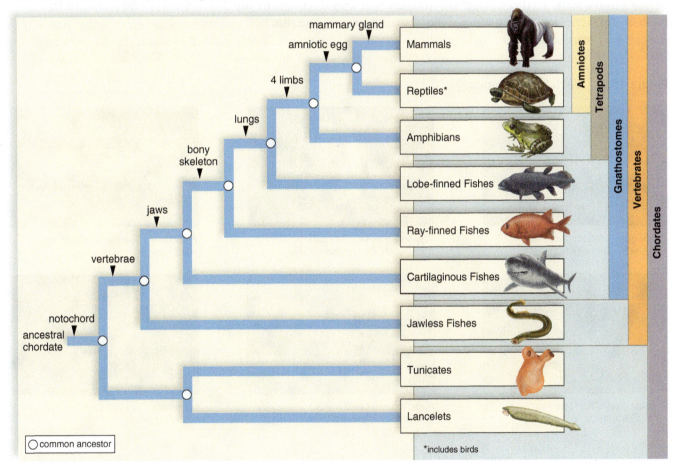

The other animal groups in Figure 16.16 are **vertebrates** in which the notochord has been replaced by the vertebral column. Vertebrates are segmented, as evidenced by their vertebral column. Fishes include three groups: The **jawless fishes** were the first to evolve, followed by the **cartilaginous fishes** and then the **bony fishes.** The bony fishes include the **ray-finned fishes** (the largest group of vertebrates) and the **lobe-finned fishes.** The first lobe-finned fishes had a bony skeleton, fleshy appendages, and a lung. These lobe-finned fishes lived in shallow pools and gave rise to the amphibians. What three innovations

called out in Figure 16.15 evolved among fishes? _____

_____

The terrestrial vertebrates are all **tetrapods** because they have four limbs. The limbs of tetrapods

are _____ appendages just like those of arthropods. Amphibians still return to the water to

reproduce, but **reptiles** are fully adapted to life on land because, among other features, they produce an **amniotic egg.** The amniotic egg is so named because the embryo is surrounded by an amniotic membrane that encloses amniotic fluid. Therefore, amniotes develop in an aquatic environment of their own making.

Do all animals develop in a water environment? _____ Explain your answer. _____

_____

In most **mammals**, including humans, the fertilized egg develops inside the female, where the unborn receives nutrients from the maternal bloodstream at the placenta.

Figure 16.16   Vertebrate groups.

Cartilaginous fishes; Caribbean reef shark, *Carcharhinus perezi*

Ray-finned fishes; Banded butterflyfish, *Chaetodon striatus*

Amphibians; Northern leopard frog, *Rana pipiens*

Reptiles; Pearl River redbelly turtle, *Pseudemys* sp.

Birds; Scissor-tailed flycatcher, *Aves tyrannidae*

Mammals; Grey fox, *Urocyon cinereoargenteus*

Laboratory Review 16

1. Name two types of organisms (not dissected) that belong to each of these phyla:

   a. Mollusca _____

   b. Arthropoda _____

   c. Chordata _____

2. Name the type of foot and location of the foot in a clam and squid. _____

   _____

3. Associate the type of foot in a clam and squid with their way of life. _____

   _____

4. Name two characteristics of all arthropods. _____

   _____

5. Name one obvious adaptation of a crayfish and one adaptation of a grasshopper to their environment.

   _____

6. How do insects assist the transport of oxygen to flight muscles? _____

   _____

7. How do you know that the grasshopper adaptation you provided in question 5 is an adaptation to life on

   land? _____

   _____

8. Name two characteristics of all chordates. _____

   _____

9. What innovation during the course of evolution can you properly associate with these groups of vertebrates?

   a. Ray-finned fishes _____

   b. Amphibians _____

   c. Reptiles _____

10. What type of skeleton do both arthropods and vertebrates share? _____

    _____

17

Basic Mammalian Anatomy I

Learning Outcomes

17.1 External Anatomy
- Compare the limbs of a pig to the limbs of a human.
- Identify the sex of a fetal pig.

17.2 Oral Cavity and Pharynx
- Find and identify the teeth, tongue, and hard and soft palates.
- Identify and state a function for the epiglottis, glottis, and esophagus.
- Name the two pathways that cross in the pharynx.

17.3 Thoracic and Abdominal Incisions
- Identify the thoracic cavity and the abdominal cavity.
- Find and identify the diaphragm.

17.4 Neck Region
- Find, identify, and state a function for the thymus gland, the larynx, and the thyroid gland.

17.5 Thoracic Cavity
- Identify the three compartments and the organs of the thoracic cavity.

17.6 Abdominal Cavity
- Find, identify, and state a function for the liver, stomach, spleen, small intestine, gallbladder, pancreas, and large intestine. Describe where these organs are positioned in relation to one another.

17.7 Human Anatomy
- Using a human torso model, find, identify, and state a function for the organs studied in this laboratory.
- Associate each organ with a particular system of the body.

Introduction

In this laboratory, you will dissect a fetal pig. Alternately your instructor may choose to have you observe a pig that has already been dissected. Both pigs and humans are mammals; therefore, you will be studying mammalian anatomy. The period of pregnancy, or gestation, in pigs is approximately 17 weeks (compared with an average of 40 weeks in humans). The piglets used in class will usually be within 1 to 2 weeks of birth.

The pigs may have a slash in the right neck region, indicating the site of blood drainage. A red latex solution may have been injected into the **arterial system,** and a blue latex solution may have been injected into the **venous system** of the pigs. If so, when a vessel appears red, it is an artery, and when a vessel appears blue, it is a vein.

As a result of this laboratory, you should gain an appreciation of which organs work together. For example, the liver and the pancreas aid the digestion of fat in the small intestine.

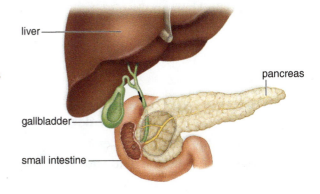

17.1 External Anatomy

Mammals are characterized by the presence of mammary glands and hair. Mammals also occur in two distinct sexes, males and females, often distinguishable by their external **genitals,** the reproductive organs.

Both pigs and humans are placental mammals, which means that development occurs within the uterus of the mother. An **umbilical cord** stretches externally between the fetal animal and the **placenta,** where carbon dioxide and organic wastes are exchanged for oxygen and organic nutrients.

Pigs and humans are tetrapods—that is, they have four limbs. Pigs walk on all four of their limbs; in fact, they walk on their toes, and their toenails have evolved into hooves. In contrast, humans walk only on the feet of their legs.

Observation: External Anatomy

Body Regions and Limbs

1. Place a pig in a dissecting pan, and observe the following body regions: the rather large head; the short, thick neck; the cylindrical trunk with two pairs of appendages (forelimbs and hindlimbs); and the short tail (Fig. 17.1a). The tail is an extension of the vertebral column.

> ⚠️ **Latex gloves** Wear protective safety goggles, latex gloves, and protective clothing when handling preserved animal organs. Exercise caution when using sharp instruments during this experiment. Wash hands thoroughly upon completion of this experiment.

2. Examine the four limbs, and feel for the joints of the digits, wrist, elbow, shoulder, hip, knee, and ankle.
3. Determine which parts of the forelimb correspond to your arm, elbow, forearm, wrist, and hand.
4. Do the same for the hindlimb, comparing it with your thigh and leg.
5. The pig walks on its toenails, which would be like a ballet dancer on "tiptoe." Notice how your heel touches the ground when you walk. Where is the heel of the pig? _____

Umbilical Cord

1. Locate the umbilical cord arising from the ventral (toward the belly) portion of the abdomen.
2. Note the cut ends of the umbilical blood vessels. If they are not easily seen, cut the umbilical cord near the end and observe this new surface.
3. What is the function of the umbilical cord? _____

_____

Nipples and Hair

1. Locate the small **nipples,** the external openings of the **mammary glands.** The nipples are *not* an indication of sex, since both males and females possess them. How many nipples does a pig have? _____

   When is it advantageous for a pig to have so many nipples? _____

_____

2. Can you find hair on the pig? _____ Where? _____

_____

Directional Terms for Dissecting Fetal Pig

Anterior: toward the head end	Ventral: toward the belly
Posterior: toward the hind end	Dorsal: toward the back

Figure 17.1 External anatomy of the fetal pig.

a. Body regions and limbs. **b, c.** The sexes can be distinguished by the external genitals.

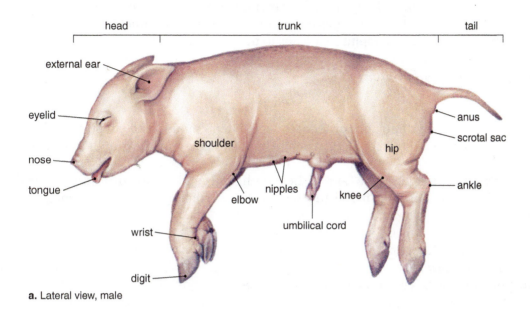

a. Lateral view, male

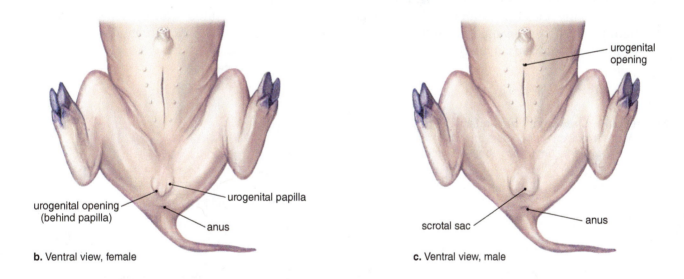

b. Ventral view, female

c. Ventral view, male

Anus and External Genitals

1. Locate the **anus** under the tail. Name the organ system that ends in the opening called the anus. _____

2. In females, locate the **urogenital opening,** just anterior to the anus, and a small, fleshy **urogenital papilla** projecting from the urogenital opening (Fig. 17.1*b*).

3. In males, locate the urogenital opening just posterior to the umbilical cord. The duct leading to it runs forward from between the legs in a long, thick tube, the **penis,** which can be felt under the skin. In males, the urinary system and the genital system are always joined (Fig. 17.1*c*).

4. You are responsible for identifying pigs of both sexes. What sex is the pig you are examining? _____

    Be sure to look at a pig of the opposite sex that another group of students is dissecting.

17.2 Oral Cavity and Pharynx

The **oral cavity** is the space in the mouth that contains the tongue and the teeth. The **pharynx** is dorsal to the oral cavity and has three openings: The glottis is an opening through which air passes on its way to the **trachea** (the windpipe) and lungs. The esophagus is a portion of the digestive tract that leads through the neck and thorax to the stomach. The **nasopharynx** leads to the nasal passages.

Observation: Oral Cavity and Pharynx

Oral Cavity

1. Insert a sturdy pair of scissors into one corner of the specimen's mouth, and cut posteriorly (toward the hind end) for approximately 4 cm. Repeat on the opposite side until the mouth is open as in Figure 17.2.
2. Place your thumb on the tongue at the front of the mouth, and gently push downward on the lower jaw. This will tear some of the tissue in the angles of the jaws so that the mouth will remain partly open (Fig. 17.2).
3. Note small, underdeveloped teeth in both the upper and lower jaws. Care should be taken, because teeth can be very sharp. Other embryonic, nonerupted teeth may also be found within the gums. The teeth are used to chew food.
4. Examine the tongue, which is partly attached to the lower jaw region but extends posteriorly and is attached to a bony structure at the back of the oral cavity (Fig. 17.2). The tongue manipulates food for swallowing.
5. Locate the hard and soft palates (Fig. 17.2). The **hard palate** is the ridged roof of the mouth that separates the oral cavity from the nasal passages. The **soft palate** is a smooth region posterior to the hard palate. An extension of the soft palate—the **uvula**—hangs down into the throat in humans. (A pig does not have a uvula.)

Figure 17.2 Oral cavity of the fetal pig.
The roof of the oral cavity contains the hard and soft palates, and the tongue lies above the floor of the oral cavity.

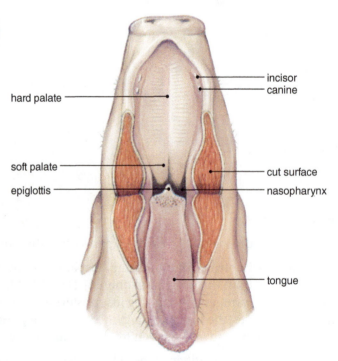

incisor
canine
hard palate
soft palate
cut surface
epiglottis
nasopharynx
tongue

Pharynx

1. Push down on the jaws until they have opened far enough to reveal a slightly pointed flap of tissue that points dorsally (toward the back) (Fig. 17.2). This flap is the **epiglottis,** which covers the glottis. The **glottis** leads to the trachea (Fig. 17.3*a*).
2. Posterior and dorsal to the glottis, find the opening into the **esophagus,** a tube that takes food to the stomach. Note the proximity of the glottis and the opening to the esophagus. Each time the pig—or a human—swallows, the epiglottis instantly closes to keep food and fluids from going into the lungs via the trachea.
3. Insert a blunt probe into the glottis, and note that it enters the trachea. Remove the probe, insert it into the esophagus, and note the position of the esophagus beneath the trachea.
4. Make two lateral cuts at the edge of the hard palate.
5. Posterior to the soft palate, locate the openings to the nasal passages.
6. Explain why it is correct to say that the air and food passages cross in the pharynx.

_____

_____

_____

Figure 17.3   Air and food passages in the fetal pig.
The air and food passages cross in the pharynx. **a.** Drawing. **b.** Dissection of specimen.

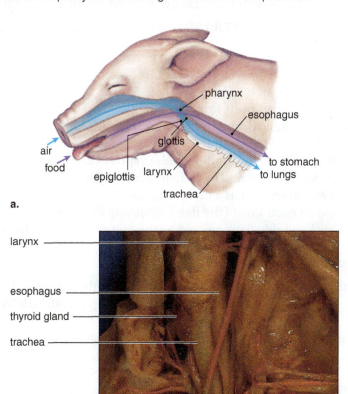

17.3 Thoracic and Abdominal Incisions

First, prepare your pig according to the following directions, and then make thoracic and abdominal incisions so that you will be able to study the internal anatomy of your pig.

Preparation of Pig for Dissection

1. Place the fetal pig on its back in the dissecting pan.
2. Tie a cord around one forelimb, and then bring the cord around underneath the pan to fasten back the other forelimb.
3. Spread the hindlimbs in the same way.
4. With scissors always pointing up (never down), make the following incisions to expose the thoracic and abdominal cavities. The incisions are numbered on Figure 17.4 to correspond with the following steps.

Thoracic Incisions

1. Cut anteriorly up from the **diaphragm,** a structure that separates the thoracic cavity from the abdominal cavity, until you reach the clump of hair below the chin.
2. Make two lateral cuts, one on each side of the midline incision anterior to the forelimbs, taking extra care not to damage the blood vessels around the heart.
3. Make two lateral cuts, one on each side of the midline just posterior to the forelimbs and anterior to the diaphragm, following the ends of the ribs. Pull back the flaps created by these cuts to expose the **thoracic cavity.** You will dissect the thoracic cavity later, but in the meantime list the organs you find in

the thoracic cavity. (See Figs. 17.5 and 17.6.) _____

_____

Abdominal Incisions

4. With scissors pointing up, cut posteriorly from the diaphragm to the umbilical cord.
5. Make a flap containing the umbilical cord by cutting a semicircle around the cord and by cutting posteriorly to the left and right of the cord.
6. Make two cuts, one on each side of the midline incision posterior to the diaphragm. Examine the diaphragm, attached to the chest wall by radially arranged muscles. The central region of the diaphragm, called the **central tendon,** is a membranous area.
7. Make two more cuts, one on each side of the flap containing the umbilical cord and just anterior to the hindlimbs. Pull back the side flaps created by these cuts to expose the **abdominal cavity.**

Complete Preparation of Pig for Dissection

1. Lifting the flap with the umbilical cord requires cutting the **umbilical vein.** Before cutting the umbilical vein, tie a thread on each side of the vein. Cut the vein but keep the threads in place for future reference.
2. As soon as you have opened the abdominal cavity, rinse out the pig. If you have a problem with excess fluid, obtain a disposable plastic pipet to suction off the liquid.

Answer These Questions

1. Name the two cavities separated by the diaphragm. _____

2. You will dissect the abdominal cavity later, but in the meantime list the organs located in the

abdominal cavity. (See Figs. 17.5 and 17.6.) _____

_____

_____

Figure 17.4 Ventral view of the fetal pig indicating incisions.
These incisions are to be made preparatory to dissecting the internal organs. They are numbered here in the order they should be done.

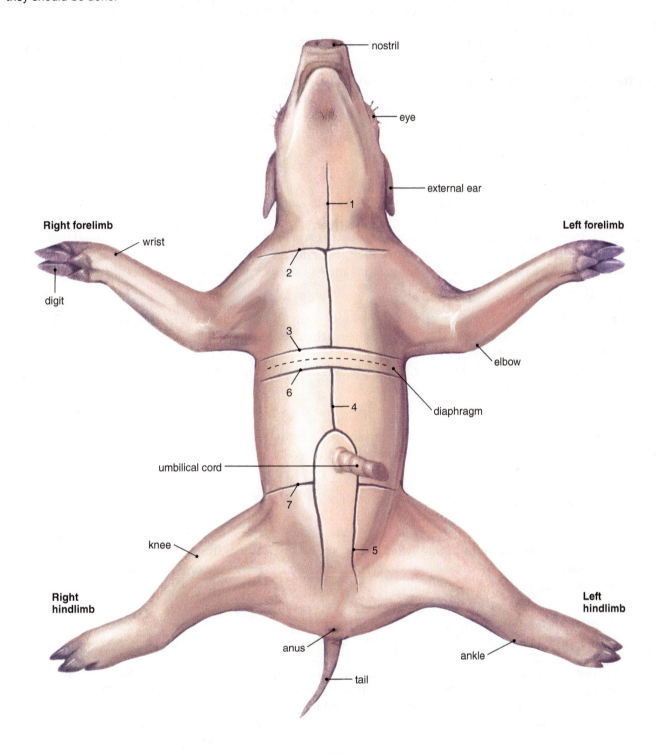

17.4 Neck Region

Several organs in the neck region are of interest. Use Figures 17.3*b* and 17.5 as a guide to locate these organs, but *keep all the flaps* in order to close the thoracic and abdominal cavities at the end of the laboratory session.

The **thymus gland** is a part of the lymphatic system. Certain white blood cells called T (for "thymus") lymphocytes mature in the thymus gland and help us fight disease. The **larynx,** or voice box, sits atop the **trachea,** or windpipe. The esophagus is a portion of the digestive tract that leads to the stomach. The **thyroid gland,** a part of the endocrine system, secretes hormones that travel in the blood and act upon other body cells. These hormones (e.g., thyroxine) regulate the rate at which metabolism occurs in cells.

Observation: Neck Region

Thymus Gland

1. Move the skin apart in the neck region just below the hairs mentioned earlier. If necessary, cut the body wall laterally to make flaps.
2. If necessary, *cut through and clear away muscle* to expose the thymus gland, a diffuse gland that lies among the muscles. Later you will notice that the thymus flanks the thyroid and overlies the heart. The thymus is particularly large in fetal pigs, since their immune systems are still developing.

Larynx, Trachea, and Esophagus

1. Probe down into the deeper layers of the neck. Medially (toward the center), beneath several strips of muscle, find the hard-walled larynx and the trachea, which are parts of the respiratory passage. Dorsal to the trachea, find the esophagus.
2. Open the mouth and insert a probe into the glottis and esophagus from the pharynx to better understand the orientation of these two organs.

Thyroid Gland

Locate the thyroid gland just posterior to the larynx, lying ventral to (on top of) the trachea. The thyroid gland secretes hormones that increase metabolism within cells.

17.5 Thoracic Cavity

As previously mentioned, the body cavity of mammals, including humans, is divided by the diaphragm into the thoracic cavity and the abdominal cavity. The heart and lungs are in the thoracic cavity (Figs. 17.5 and 17.6). The **heart** is a pump for the cardiovascular system, and the **lungs** are organs of the respiratory system where gas exchange occurs.

Observation: Thoracic Cavity

Heart and Lungs

1. In order to fold back the chest wall flaps, tear the thin membranes that divide the thoracic cavity into three compartments. The three compartments are the **left pleural cavity** containing the left lung, the **right pleural cavity** containing the right lung, and the **pericardial cavity** containing the heart.
2. Examine the lungs. Locate the four lobes of the right lung and the three lobes of the left lung. The trachea, dorsal to the heart, divides into the **bronchi,** which enter the lungs.
3. Trace the path of air from the nasal passages to the lungs.

_____

_____

Figure 17.5 Internal anatomy of the fetal pig.

The major organs are featured in this drawing. In the fetal pig, a vessel colored red is an artery, and a vessel colored blue is a vein. (The color does not indicate whether this vessel carries O_2-rich or O_2-poor blood.) Contrary to this drawing, do not cut the flaps, because they can be closed to protect the thoracic and abdominal cavities.

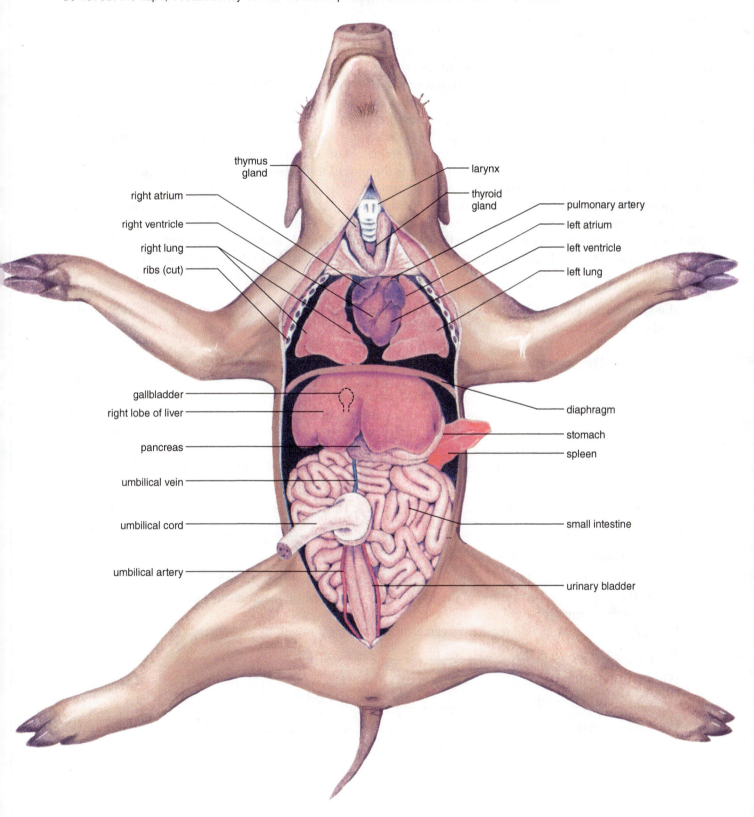

thymus gland

larynx

right atrium

thyroid gland

pulmonary artery

right ventricle

left atrium

right lung

left ventricle

ribs (cut)

left lung

gallbladder

diaphragm

right lobe of liver

stomach

pancreas

spleen

umbilical vein

umbilical cord

small intestine

umbilical artery

urinary bladder

17.6 Abdominal Cavity

The abdominal wall and organs are lined by a membrane called **peritoneum**, consisting of epithelium supported by connective tissue. Double-layered sheets of peritoneum, called **mesenteries**, project from the body wall and support the organs.

The **liver**, the largest organ in the abdomen (Fig. 17.6), performs numerous vital functions, including (1) disposing of worn-out red blood cells, (2) producing bile, (3) storing glycogen, (4) maintaining the blood glucose level, and (5) producing blood proteins.

The abdominal cavity also contains organs of the digestive tract, such as the stomach, small intestine, and large intestine. The **stomach** (see Fig. 17.5) stores food and has numerous gastric glands that secrete gastric juice, which digests protein. The **small intestine** is the part of the digestive tract that receives secretions from the pancreas and gallbladder. Besides being an area for the digestion of all components of food—carbohydrate, protein, and fat—the small intestine absorbs the products of digestion: glucose, amino acids, glycerol, and fatty acids. The **large intestine** is the part of the digestive tract that absorbs water and prepares feces for defecation at the anus.

The **gallbladder** stores and releases bile, which aids the digestion of fat. The **pancreas** (see Fig. 17.5) is both an exocrine and an endocrine gland. As an exocrine gland, it produces and secretes pancreatic juice, which digests all the components of food in the small intestine. Both bile and pancreatic juice enter the duodenum (the first, straight part of the small intestine) by way of ducts. As an endocrine gland, the pancreas secretes the hormones insulin and glucagon into the bloodstream. Insulin and glucagon regulate blood glucose levels.

The **spleen** (see Fig. 17.5) is a lymphoid organ in the lymphatic system that contains both white and red blood cells. It purifies blood and disposes of worn-out red blood cells.

Observation: Abdominal Cavity

Liver

1. If your pig is partially filled with dark, brownish material, take your animal to the sink and rinse it out. This material is clotted blood. Consult your instructor before removing any red or blue latex masses, since they may enclose organs you will need to study.
2. Locate the liver, a large, brown organ. Its anterior surface is smoothly convex and fits snugly into the concavity of the diaphragm.
3. Name several functions of the liver. _____

_____

Stomach and Spleen

1. Push aside and identify the stomach, a large sac dorsal to the liver on the left side.
2. Locate the point near the midline of the body where the **esophagus** penetrates the diaphragm and joins the stomach.
3. Find the spleen, a long, flat, reddish organ attached to the stomach by mesentery.
4. The stomach is a part of the _____ system.

   What is its function? _____
5. The spleen is a part of the _____ system.

   What is its function? _____

Figure 17.6 Internal anatomy of the fetal pig.

Most of the major organs are shown in this photograph. The stomach has been removed. The spleen, gallbladder, and pancreas are not visible. *Do not* remove any organs or flaps from your pig.

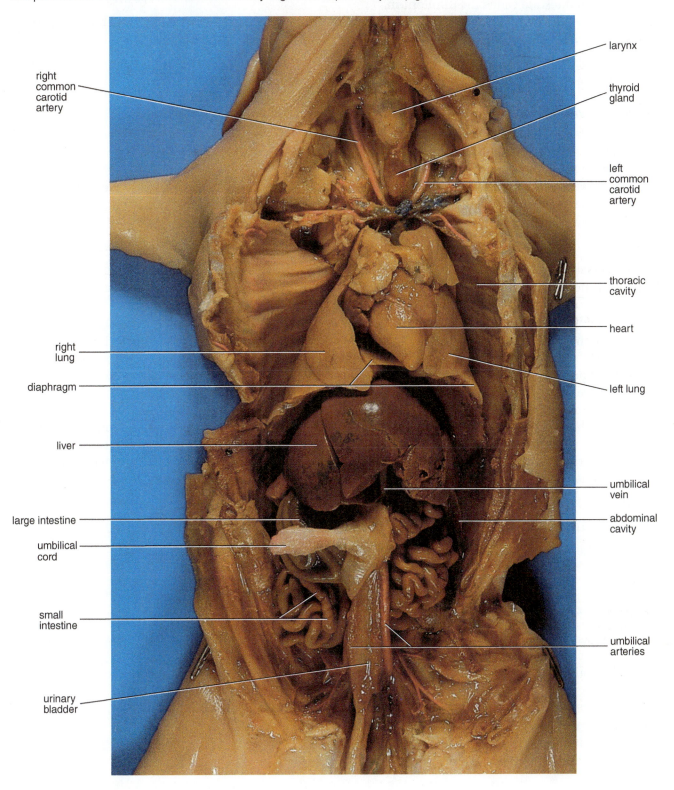

Small Intestine

1. Look posteriorly where the stomach makes a curve to the right and narrows to join the anterior end of the small intestine called the **duodenum.**
2. From the duodenum, the small intestine runs posteriorly for a short distance and is then thrown into an irregular mass of bends and coils held together by a common mesentery.
3. The small intestine is a part of the _____ system.

   What is its function? _____

Gallbladder and Pancreas

1. Locate the **bile duct,** which runs in the mesentery stretching between the liver and the duodenum. Find the gallbladder, embedded in the liver on the underside of the right lobe. It is a small, greenish sac.
2. Lift the stomach and locate the pancreas, the light-colored, diffuse gland lying in the mesentery between the stomach and the small intestine. The pancreas has a duct that empties into the duodenum of the small intestine.
3. What is the function of the gallbladder? _____
4. What is the function of the pancreas? _____

   _____

Large Intestine

1. Locate the distal (far) end of the small intestine, which joins the large intestine posteriorly, in the left side of the abdominal cavity (right side in humans). At this junction, note the **cecum,** a blind pouch.
2. Compare the large intestine of a pig to Figure 17.7. The organ does not have the same appearance in humans.
3. Follow the main portion of the large intestine, known as the **colon,** as it runs from the point of juncture with the small intestine into a tight coil (spiral colon), then out of the coil anteriorly, then posteriorly again along the midline of the dorsal wall of the abdominal cavity. In the pelvic region, the **rectum** is the last portion of the large intestine. The rectum leads to the **anus.**
4. The large intestine is a part of the _____ system.
5. What is the function of the large intestine? _____
6. Trace the path of food from the mouth to the anus. _____

   _____

Storage of Pigs

1. Before leaving the laboratory, place your pig in the plastic bag provided.
2. Expel excess air from the bag, and tie it shut.
3. Write your name and section on the tag provided, and attach it to the bag. Your instructor will indicate where the bags are to be stored until the next laboratory period.
4. Clean the dissecting tray and tools, and return them to their proper location.
5. Wipe off your goggles.
6. Wash your hands.

17.7 Human Anatomy

Humans and pigs are both mammals, and their organs are similar. A human torso model shows the exact location of the organs in humans (Fig. 17.7). Learn to associate these organs with their particular system. Six systems are color-coded in Figure 17.7.

Figure 17.7 Human internal organs.

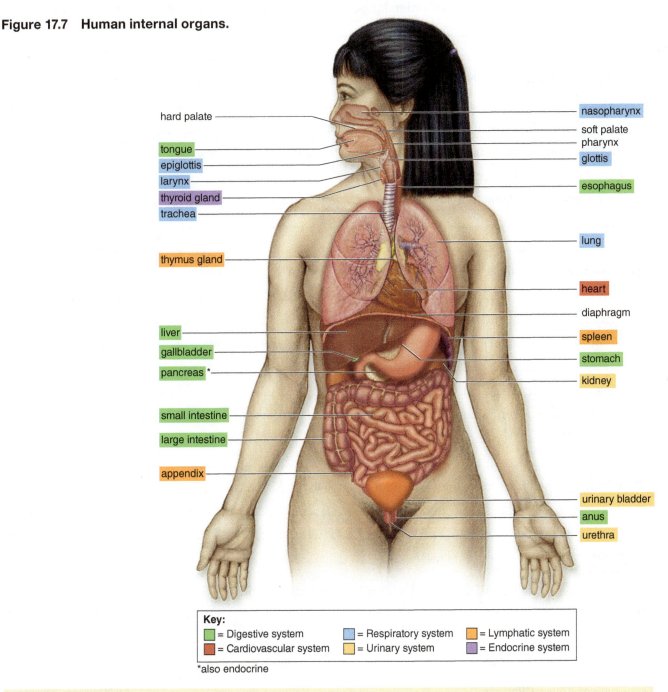

Key:
- ■ = Digestive system
- ■ = Respiratory system
- ■ = Lymphatic system
- ■ = Cardiovascular system
- ■ = Urinary system
- ■ = Endocrine system

*also endocrine

Observation: Human Torso

1. Examine a human torso model, and using Figure 17.7 as a guide, locate the same organs just dissected in the fetal pig.
2. Name any observed major differences between pig internal anatomy and human internal anatomy.

1. What two features indicate that a pig is a mammal? _____
   _____

2. Put the following organs in logical order: lungs, nasal passages, nasopharynx, trachea, bronchi, glottis.
   _____
   _____

3. What difficulty would probably arise if a person were born without an epiglottis? _____
   _____
   _____

4. The embryonic coelom may be associated with what two cavities studied in this laboratory? _____
   _____

5. Name two principal organs in the thoracic cavity, and give a function for each. _____
   _____
   _____

6. What difficulty would arise if a person were born without a thymus gland? _____
   _____
   _____

7. Name the largest organ in the abdominal cavity, and list several functions. _____
   _____
   _____

8. A large portion of the abdominal cavity is taken up with digestive organs. What are they? _____
   _____
   _____

9. Why is it proper to associate the gallbladder with the liver? _____
   _____
   _____

10. Where would you find the pancreas? _____
    _____
    _____

18

Chemical Aspects of Digestion

Learning Outcomes

Introduction
- Sequence the organs of the digestive tract from the mouth to the anus.
- State the contribution of each organ, if any, to the process of chemical digestion.

18.1 Protein Digestion by Pepsin
- Associate the enzyme pepsin with the ability of the stomach to digest protein.
- Explain why stomach contents are acidic and how a warm body temperature aids digestion.

18.2 Fat Digestion by Pancreatic Lipase
- Associate the enzyme lipase with the ability of the small intestine to digest fat.
- Explain why the emulsification process assists the action of lipase.
- Explain why a change in pH indicates that fat digestion has occurred.
- Explain the relationship between time and enzyme activity.

18.3 Starch Digestion by Pancreatic Amylase
- Associate the enzyme pancreatic amylase with the ability of the small intestine to digest starch.

18.4 Requirements for Digestion
- Assuming a specific enzyme, list four factors that can affect the activity of all enzymes.
- Explain why the operative procedure that reduces the size of the stomach causes an individual to lose weight.

Introduction

In Laboratory 17, you examined the organs of digestion in the fetal pig. Now we wish to further our knowledge of the digestive process by associating certain digestive enyzmes with particular organs, as shown in Figure 18.1. This laboratory will also give us an opportunity to study the action of enzymes, much as William Beaumont did when he removed food samples through a hole in the stomach wall of his patient, Alexis St. Martin. Every few hours, Beaumont would see how well the food had been digested.

In Laboratory 5 we learned that enzymes are very specific and usually participate in only one type of reaction. The active site of an enzyme has a shape that accommodates its substrate, and if an environmental factor such as a boiling temperature or a wrong pH alters this shape, the enzyme loses its ability to function well, if at all. We will have an opportunity to make these observations with controlled experiments. The box on the next page reviews what is meant by a controlled experiment.

> 🕐 **Planning Ahead** Be advised that protein digestion (page 227) requires 1½ hours and fat digestion (page 229) requires 1 hour. Also, a boiling water bath is required for starch digestion (page 231).

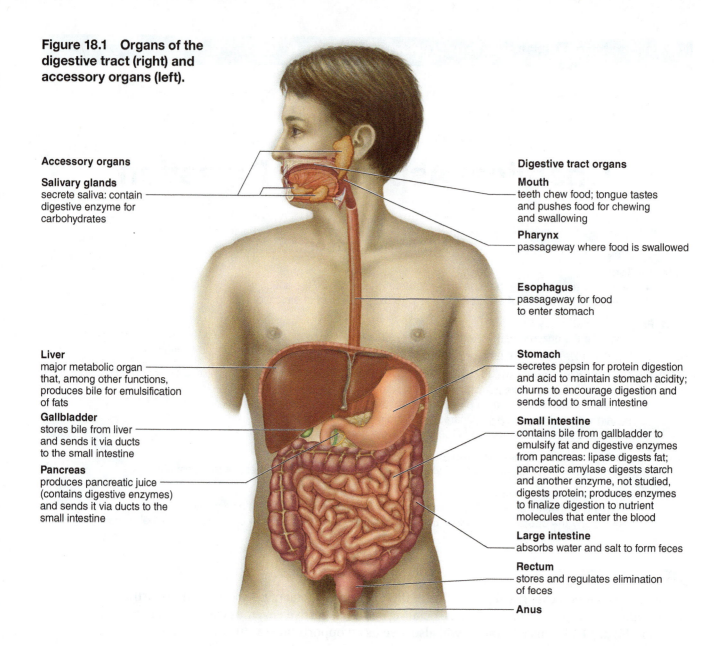

Figure 18.1 Organs of the digestive tract (right) and accessory organs (left).

Accessory organs

Salivary glands
secrete saliva: contain digestive enzyme for carbohydrates

Liver
major metabolic organ that, among other functions, produces bile for emulsification of fats

Gallbladder
stores bile from liver and sends it via ducts to the small intestine

Pancreas
produces pancreatic juice (contains digestive enzymes) and sends it via ducts to the small intestine

Digestive tract organs

Mouth
teeth chew food; tongue tastes and pushes food for chewing and swallowing

Pharynx
passageway where food is swallowed

Esophagus
passageway for food to enter stomach

Stomach
secretes pepsin for protein digestion and acid to maintain stomach acidity; churns to encourage digestion and sends food to small intestine

Small intestine
contains bile from gallbladder to emulsify fat and digestive enzymes from pancreas: lipase digests fat; pancreatic amylase digests starch and another enzyme, not studied, digests protein; produces enzymes to finalize digestion to nutrient molecules that enter the blood

Large intestine
absorbs water and salt to form feces

Rectum
stores and regulates elimination of feces

Anus

What Is a Control?

The experiments in today's laboratory have both a positive control and a negative control, *which should be saved for comparison purposes until the experiment is complete*. The **positive control** goes through all the steps of the experiment and does contain the substance being tested. Therefore, positive results are expected. The **negative control** goes through all the steps of the experiment, except it does not contain the substance being tested. Therefore, negative results are expected.

For example, if a test tube contains glucose (the substance being tested) and Benedict's reagent (blue) is added, a red color develops upon heating. This test tube is the positive control; it tests positive for glucose. If a test tube does not contain glucose and Benedict's reagent is added, Benedict's is expected to remain blue. This test tube is the negative control; it tests negative for glucose.

What benefit is a positive control? Positive controls give you a standard by which to tell if the substance being tested is present (or acting properly) in an unknown sample. Negative controls ensure that the experiment is giving reliable results; after all, if a negative control should happen to give a positive result, then the entire experiment may be faulty and unreliable.

18.1 Protein Digestion by Pepsin

Certain foods, such as meat and egg whites, are rich in protein. Egg whites contain albumin, which is the protein used in this Experimental Procedure. Protein is digested by **pepsin** in the stomach (Fig. 18.2), a process described by the following reaction:

$$\text{protein} + \text{water} \xrightarrow{\text{pepsin (enzyme)}} \text{peptides}$$

The stomach has a very low pH. Does this indicate that pepsin works effectively in an acidic or a basic environment? _____ This is the pH that allows the enzyme to maintain its normal shape so that it will combine with the substrate. A warm temperature causes molecules to move about more rapidly and increases the encounters between enzyme and substrate. Therefore, you would hypothesize that the yield from this enzymatic reaction will be higher if the

pH is _____ and the temperature is _____ (body temperature 37°C).

Test for Protein Digestion

Biuret reagent is used to test for protein digestion. If digestion has not occurred, biuret reagent turns purple, indicating that protein is present. If digestion has occurred, biuret reagent turns pinkish-purple, indicating that peptides are present.

> ⚠ **Biuret reagent**   is highly corrosive. Exercise care in using this chemical. If any should spill on your skin, wash the area with mild soap and water. Follow your instructor's directions for its disposal.

Experimental Procedure: Protein Digestion

1. Label four clean test tubes (1 to 4). Using a graduated transfer pipet, add 2 ml of the albumin solution to all tubes. Albumin is a protein.
2. Add 2 ml of the pepsin solution to tubes 1 to 3, as listed in Table 18.1.
3. Add 2 ml of 0.2% HCl to tubes 1 and 2. HCl simulates the acidic conditions of the stomach.
4. Add 2 ml of water to tube 3 and 4 ml of water to tube 4, as listed in Table 18.1.
5. Swirl to mix the tubes. Tube 2 remains at room temperature, but the other three are incubated for 1½ hours. Record the temperature for each tube in Table 18.1.
6. Remove the tubes from the incubator and place all four tubes in a tube rack. Add 2 ml of biuret reagent to all tubes and observe. Record your results in Table 18.1 as + or – to indicate digestion or no digestion.

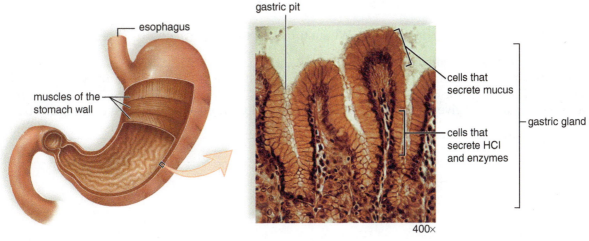

Figure 18.2 Digestion of protein.
Pepsin, produced by the gastric glands of the stomach, helps digest protein.

Table 18.1 Protein Digestion by Pepsin

Tube	Contents	Temperature	Digestion (+ or −)	Explanation
1	Albumin Pepsin HCl Biuret reagent			
2	Albumin Pepsin HCl Biuret reagent			
3	Albumin Pepsin Water Biuret reagent			
4	Albumin Water Biuret reagent			

Conclusions: Protein Digestion

- Explain your results in Table 18.1 by reasoning why digestion did or did not occur.
  To be complete, consider all the requirements for an enzymatic reaction as listed in Table 18.4. Now
  show here that tube 1 met all the requirements for digestion:

  Pepsin is the correct _____.

  Albumin is the correct _____.

  37°C is the optimum _____.

  HCl provides the optimum _____.

  1½ hours provide _____ for the reaction to occur.

- Review "What Is a Control?" on page 226. Which tube was the negative control? _____

  Explain why it was the negative control. _____

  _____

- If this control tube had given a positive result for protein digestion, what could you conclude about

  this experiment? _____

  _____

18.2 Fat Digestion by Pancreatic Lipase

Lipids include fats (e.g., butterfat) and oils (e.g., sunflower, corn, olive, and canola). Lipids are digested by **pancreatic lipase** in the small intestine (Fig. 18.3).

Figure 18.3 Emulsification and digestion of fat.
Bile from the liver (stored in the gallbladder) enters the small intestine, where lipase in pancreatic juice from the pancreas digests fat.

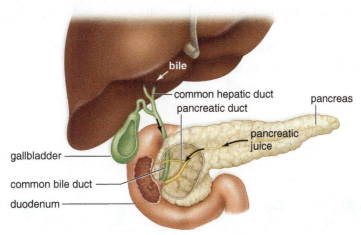

The following two reactions describe fat digestion:

1.
$$\text{fat} \xrightarrow{\text{bile (emulsifier)}} \text{fat droplets}$$

2.
$$\text{fat droplets} + \text{water} \xrightarrow{\text{lipase (enzyme)}} \text{glycerol} + \text{fatty acids}$$

With regard to the first step, consider that fat is not soluble in water, yet lipase makes use of water when it digests fat. Therefore, bile is needed to emulsify fat—cause it to break up into fat droplets that disperse in water. The reason for dispersal is that bile contains molecules with two ends. One end is soluble in fat, and the other end is soluble in water. Bile can emulsify fat because of this.

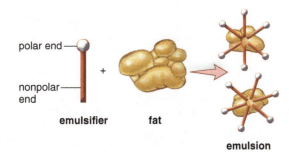

With regard to the second step, would the pH of the solution be lower before or after the enzymatic reaction? (*Hint:* Remember that an acid decreases pH and a base increases pH.) _____

_____

Test for Fat Digestion

In the test for fat digestion, you will be using a pH indicator, which changes color as the solution in the test tube goes from basic conditions to acidic conditions. Phenol red is a pH indicator that is red in basic solutions and yellow in acidic solutions.

Experimental Procedure: Fat Digestion

1. Label three clean test tubes (1 to 3). Using a graduated transfer pipet, add 1 ml of vegetable oil to all tubes.
2. Add 2 ml of phenol red solution to each tube. What role does phenol red play? _____
3. Add 2 ml of pancreatic lipase (pancreatin) to tubes 1 and 2 and 2 ml of water to tube 3, as listed in Table 18.2. What role does lipase play? _____
4. Add a pinch of bile salts to tube 1.
5. Record the initial color of all tubes in Table 18.2.
6. Incubate all three tubes at 37°C and check every 20 minutes.
7. Record any color change and how long it took to see this color change in Table 18.2.

Table 18.2 Fat Digestion by Pancreatic Lipase

Tube	Contents	Color		Time Taken	Explanation
		Initial	*Final*		
1	Vegetable oil Phenol red Pancreatin Bile salts				
2	Vegetable oil Phenol red Pancreatin				
3	Vegetable oil Phenol red Water				

Conclusions: Fat Digestion

- Explain your results in Table 18.2 by reasoning why digestion did or did not occur.
- What role did bile salts play in this experiment? _____
  _____
- What role did phenol red play in this experiment? _____
- Review "What Is a Control?" on page 226. Which test tube in this experiment could be considered a negative control? _____
  _____

18.3 Starch Digestion by Pancreatic Amylase

Starch is present in bakery products and in potatoes, rice, and corn. Starch is digested by **pancreatic amylase** in the small intestine, a process described by the following reaction:

$$\text{starch} + \text{water} \xrightarrow{\text{amylase (enzyme)}} \text{maltose}$$

1. If digestion *does not* occur, which will be present—starch or maltose? _____

2. If digestion *does* occur, which will be present—starch or maltose? _____

Tests for Starch Digestion

You will be using two tests for starch digestion:

1. If digestion has not taken place, the iodine test for starch will be positive (+). If digestion has occurred, the iodine test for starch will be negative (−).
2. If digestion of starch has taken place, the Benedict's test for sugar (maltose) will be positive (+). If digestion has not taken place, the Benedict's test for sugar will be negative (−).

> ⚠️ **Benedict's reagent** is highly corrosive. Use protective eyewear when performing this experiment. Exercise care in using this chemical. If any should spill on your skin, wash the area with mild soap and water. Follow your instructor's directions for disposal of this chemical.

To test for sugar, add five drops of Benedict's reagent. Place the tube in a boiling water bath for a few minutes, and note any color changes. Boiling the test tube is necessary for the Benedict's reagent to react. The color change varies from red (high concentration of glucose) to orange, yellow, and green (low concentration of glucose). No color change means an absence of glucose.

Experimental Procedure: Starch Digestion

1. Label six clean test tubes (1 to 6).
2. Using a graduated transfer pipet, add l ml of pancreatic-amylase solution to tubes 1 to 4 and 1 ml of water to tubes 5 and 6.
3. Test tubes 1 and 2 immediately.

   Tube 1 Shake the starch solution and add 1 ml of starch solution. Immediately add five drops of iodine to test for starch. Put this tube in a test tube rack and record your results in Table 18.3.

   Tube 2 Shake the starch solution and add 1 ml of starch solution. Immediately add five drops of Benedict's reagent, and place the tube in a boiling water bath to test for sugar. Put this tube in the test tube rack and record your results in Table 18.3.

4. Shake the starch suspension and add 1 ml of starch suspension to tubes 3 to 6. Allow the tubes to stand for 30 minutes.

   Tubes 3 and 5 After 30 minutes, test for starch using the iodine test. Place these tubes in the test tube rack and record your results in Table 18.3.

   Tubes 4 and 6 After 30 minutes, test for sugar using the Benedict's test. Place these tubes in the test tube rack and record your results in Table 18.3.

5. Examine all your tubes in the test rube rack and decide whether digestion occurred (+) or did not occur (−). Complete Table 18.3.

Table 18.3   Starch Digestion by Amylase

Tube	Contents	Time*	Type of Test	Test Results (+ or −)	Digestion (+ or −)
1	Pancreatic amylase Starch	0	Iodine	+	
2	Pancreatic amylase Starch				
3	Pancreatic amylase Starch				
4	Pancreatic amylase Starch				
5	Water Starch				
6	Water Starch				

* Enter either 0 for immediately or T for after 30 minutes.

Conclusions: Starch Digestion

- Considering tubes 1 and 2, this Experimental Procedure showed that _____ must pass for digestion to occur.
- Considering tubes 5 and 6, this Experimental Procedure showed that an active _____ must be present for digestion to occur.
- Why would you not recommend doing the test for starch and the test for sugar on the same tube? _____

  _____

- Which test tubes served as a negative control for this experiment? _____

  Explain your answer. _____

Absorption of Sugars and Other Nutrients

Figure 18.4 shows that the folded lining of the small intestine has many fingerlike projections called villi. The small intestine not only digests food; it also absorbs the products of digestion, such as sugars from carbohydrate digestion, amino acids from protein digestion, and glycerol and fatty acids from fat digestion at the villi.

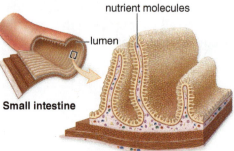

Figure 18.4   Anatomy of the small intestine.
Nutrients enter the bloodstream across the much-convoluted walls of the small intestine.

18.4 Requirements for Digestion

Explain in Table 18.4 how each of the requirements listed influences effective digestion.

Table 18.4 Requirements for Digestion	
Requirement	**Explanation**
Specific enzyme	
Specific substrate	
Warm temperature	
Specific pH	
Time	
Fat emulsifier	

To lose weight, some obese individuals undergo an operation in which (1) the stomach is reduced to the size of a golf ball and (2) food bypasses the duodenum (first 2 feet) of the intestine. Answer these questions to explain how this operation would affect the requirements for digestion.

1. How is the amount of substrate reduced? _____

2. How is the amount of digestive enzymes reduced? _____

3. How is time reduced? _____

4. What makes the pH of the small intestine higher than before? _____

5. How is fat emulsification reduced? _____

6. How does surgery to reduce obesity sometimes result in malnutrition? _____

Laboratory Review 18

1. Where in the body does starch digestion occur? _____ and _____ Protein

   digestion occur? _____ and _____ Fat digestion occur? _____

2. Why would you not expect amylase to digest protein? _____

   _____

3. Relate the expectation of more product per length of time to the fact that enzymes are used over and over.

   _____

   _____

4. Why do enzymes work better at their optimum pH? _____

   _____

5. Why is an emulsifier needed for the lipase experiment but not for the pepsin and amylase experiments?

   _____

6. Which of the following two combinations is most likely to result in digestion?

   a. Pepsin, protein, water, body temperature

   b. Pepsin, protein, hydrochloric acid (HCl), body temperature

   Explain your answer. _____

7. Which of the following two combinations is most likely to result in digestion?

   a. Amylase, starch, water, body temperature, testing immediately

   b. Amylase, starch, water, body temperature, waiting 30 minutes

   Explain your answer. _____

   _____

8. Relate the composition of fat to the test used for fat digestion. _____

   _____

   _____

9. Given that, in this laboratory, you tested for the action of digestive enzymes on their substrates, what

   substance would be missing from a negative control sample? _____

   _____

   _____

10. What substance would be present in a positive control sample? _____

    _____

19

Energy Requirements and Ideal Weight

Learning Outcomes

19.1 Average Daily Energy Intake
- Keep a food diary and use it to calculate your average daily energy intake.

19.2 Average Daily Energy Requirement
- Keep a physical activity diary and use it to calculate average daily energy required for physical activity.
- Calculate the daily energy required for your basal metabolism.
- Calculate the daily energy required for your specific dynamic action.
- Calculate your total average daily energy requirement.

19.3 Comparison of Average Daily Energy Intake and Average Daily Energy Requirement
- Predict whether weight gain or weight loss will occur after energy intake is compared to energy required.

19.4 Ideal Weight
- Consult height and weight tables to determine ideal weight.
- Calculate ideal weight based on body mass index (BMI).
- Calculate ideal weight based on body composition.
- Make a recommendation regarding daily energy intake and daily energy required for achieving or maintaining ideal weight.

Introduction

This laboratory helps you determine your average daily energy (kcal) requirements, your average daily energy (kcal) intake, and your ideal weight. In addition to the energy content of food, other nutritional aspects are also important. Minerals and vitamins, along with specific types of biological molecules, contribute to a healthy diet. So, while the focus of this lab is on energy in our food, there is more to diet than just calories.

Before you come to the laboratory, it is necessary for you to keep two daily diaries for three days. The physical activity diary is a record of the energy you expend for physical activity. Physical activity is only a portion of your daily energy requirement. You also need energy for **basal metabolism** (energy needed when the body is resting) and **specific dynamic action** (energy needed to process food), also known as the thermal effect of food.

The food diary is a record of the energy you take in as food. If the energy you expend and the energy you take in are in balance, you will neither gain nor lose weight. If your ideal weight is different from your present weight, you will be able to determine whether to increase or decrease your physical activity and food intake to attain an ideal weight.

19.1 Average Daily Energy Intake

To complete this laboratory, you must calculate an average daily energy intake. The first method given is preferred, because it makes use of your actual intake. However, when appropriate, a hypothetical method is also provided.

Personal Diet

Follow these directions.

1. Keep a written food diary of your own design for three days. Your diary should list all food and beverages consumed throughout an entire three-day period, including snacks and alcoholic beverages.
2. Make sure your food diary contains a column for recording the energy in kcal (kilocalories)* of the food you eat. Table 19.1 may be helpful in this regard. Also, food packages often contain kcal information, as do various calorie books and computer programs. For example, after eating two slices of bread, your diary might look like this:

Day of Week _____			
Type of Food	**Portion Size**	**kcal**	**Total kcal**
Bread	Two slices	70/one slice	140

Continue in this manner until you have listed all the foods you eat in one day. Do the same for two more days.

3. Add up your total kcal intake for each day, and record the information here:

   Day 1 energy intake _____ kcal

   Day 2 energy intake _____ kcal

   Day 3 energy intake _____ kcal

   Total _____ kcal

4. Divide the total by 3 to calculate the average daily energy intake.

   Av. energy intake/day = _____ kcal

*The amount of heat required to raise 1 kg (kilogram) of water 1°C.

Table 19.1 Nutrient and Energy Content of Foods*

Breads

1 slice bread or any of the following:

¾ cup ready-to-eat cereal

⅓ cup corn

1 small potato

(1 bread = 15 g carbohydrate, 2 g protein, and 70 kcal)

Milk

1 cup skim milk or:	(2% milk—add 5 g fat)
1 cup skim-milk yogurt, plain	(whole milk—add 8 g fat)

1 cup buttermilk

½ cup evaporated skim milk or milk dessert

(1 milk = 12 g carbohydrate, 8 g protein, and 80 kcal)

Vegetables

½ cup greens

½ cup carrots

½ cup beets

(1 vegetable = 5 g carbohydrate, 1 g protein, and 25 kcal)

Fruits

½ small banana or

1 small apple

½ cup orange juice or ½ grapefruit

(1 fruit = 10 g carbohydrate and 40 kcal)

Meats (lean)

1 oz. lean meat or:

1 oz. chicken meat without the skin

1 oz. any fish

¼ cup canned tuna or 1 oz. low-fat cheese

(1 oz. low-fat meat = 7 g protein, 3 g fat, and 55 kcal)

Meats (medium fat)

1 oz. pork loin

1 egg

¼ cup creamed cottage cheese

(1 medium-fat meat = 7 g protein,

5½ g fat, and about 80 kcal)

Meats (high fat)

1 oz. high-fat meat is like:

1 oz. country-style ham

1 oz. cheddar cheese

small hot dog (frankfurter)

(1 high-fat meat = 7 g protein, 8 g fat, and 100 kcal)

Peanut butter

Peanut butter is like a meat in terms of its protein content but is very high in fat. It is estimated as (2 tbsp peanut butter = 7 g protein, 15½ g fat, and about 170 kcal)

Fats

1 tsp butter or margarine

1 tsp any oil

1 tbsp salad dressing

1 strip crisp bacon

5 small olives

10 whole Virginia peanuts

(1 fat = 5 g fat and 45 kcal)

Legumes (beans and peas)

Legumes are like meats because they are rich in protein and iron but are lower in fat than meat. They contain much starch. They can be treated as (½ cup legumes = 15 g carbohydrate, 9 g protein, 3 g fat, and 125 kcal).

Miscellaneous Foods

	Protein	Fat	Carbohydrate	kcal		Protein	Fat	Carbohydrate	kcal
	g (grams)			*kcal*		*g (grams)*			*kcal*
Ice cream (1 cup)	5	14	32	274	Beer (1 can)	1	0	14	60
Cake (1 piece)	3	1	32	149	Soft drink	0	0	37	148
Doughnuts (1)	3	11	22	199	Soup (1 cup)	2	3	15	95
Pie	3	15	51	351	Coffee and tea	0	0	0	0
Caramel candy (1 oz.)	1	3	22	119					

*Use this table for the food diary mentioned on p. 236.

Hypothetical Diet

Suppose you have decided to eat your meals at the fast-food restaurants listed in the table.

1. Circle the restaurant and the items you have chosen for breakfast, lunch, and dinner. Choose a different restaurant for each meal.
2. Record the number of kcal for the items chosen.
3. Add the number of kcal.
4. This will be your average energy intake/day.

Fast-Food Menus and kcal

Restaurant/Menu	kcal	Restaurant/Menu	kcal
Taco Bell (www.tacobell.com)		**McDonald's (www.mcdonalds.com)**	
Steak Baja chalupa	390	Egg McMuffin	300
Soft beef taco supreme	240	Sausage McMuffin w/egg	450
Chicken Baja gordita	320	Ham and egg cheese bagel	550
Spicy chicken soft taco	170	Hash browns	130
Nachos Bellgrande	760	Hotcakes (w/margarine & syrup)	350
Soft drink (small)	150	Breakfast burrito	290
Subway (www.subway.com)		6 pc. chicken nugget	280
Meatball marinara	560	Barbeque sauce	50
6" roasted chicken breast	310	Hamburger	250
6" roast beef	290	Quarter pounder w/cheese	510
Soft drink (small)	150	Big Mac	540
Burger King (www.burgerking.com)		Filet-O-Fish	380
Croissan'wich w/sausage, egg & cheese	520	Chicken McGrill w/mayo	420
Croissan'wich w/egg & cheese	320	Med. french fries	380
Egg'wich™ w/Canadian bacon & egg	380	Large french fries	500
French toast sticks	390	Hot fudge sundae	340
Lg. hash brown rounds	390	Vanilla milk shake	570
Double Whopper w/cheese and mayo	1150	Soft drink (small)	150
Whopper w/cheese and mayo	760	**Papa John's (www.papajohns.com)**	
BK big fish	710	Cheese pizza	283
King-sized fries w/salt	600	All-meat pizza	390
Med.-sized fries w/salt	360	Garden special pizza	280
King-sized onion rings	550	The works pizza	342
Med.-sized onion rings	320	Cheese sticks	180
Dutch apple pie	340	Breadsticks	140
Hershey's sundae pie	310	Garlic sauce	235
Soft drink (small)	150	Pizza sauce	25
Vanilla milk shake (medium)	720	Nacho cheese sauce	60
		Soft drink (small)	150

Av. energy intake/day = _____ kcal

19.2 Average Daily Energy Requirement

As stated at the beginning of this laboratory, the body needs energy for three purposes: (1) energy for physical activity, (2) energy to support basal metabolism, and (3) energy for the specific dynamic action of processing food. We will calculate the daily energy requirement for each of these.

Calculating Average Daily Energy Required for Physical Activity

You will need to know your average daily energy requirement for physical activity before coming to this laboratory.

1. Keep a physical activity diary for three days. Make three copies of Table 19.2, one for each day.
2. Use Scale 3 (left and right sides) in Figure 19.1, p. 241, to convert your body weight to kg (kilograms). Place this value on the appropriate line in your physical activity diary (Table 19.2).
3. Consult Table 19.3 and fill in column 4 of your physical activity diary.
4. Fill in column 5 of your physical activity diary by using a calculator and this formula:

$$\text{time spent} \times \text{energy cost} \times \text{body weight} = \text{total energy expended}$$
$$\text{(min)} \qquad\qquad\qquad \text{(kg)} \qquad\qquad \text{(kcal)}$$

Therefore, if 100 minutes are spent doing word processing, and the body weight is 77.3 kg, the kcal expended for typing is

$$100 \text{ min} \times 0.01 \text{ kcal/kg min} \times 77.3 \text{ kg} = 77.3 \text{ kcal}$$

5. Total the number of kcal expended for each of the three days, and divide this total by 3 to get the average (av.) daily energy required for physical activity:

Energy for physical activity (day 1)   _____ kcal (see p. 240)

Energy for physical activity (day 2)   _____ kcal

Energy for physical activity (day 3)   _____ kcal

Total energy for physical activity/three days   _____ kcal

Av. energy required for physical activity/day   _____ kcal

A completed food diary and physical activity diary are necessary before you begin the laboratory.

Table 19.2 Physical Activity Diary for One Day

Time of Day	Activity	Time Spent (min)	Factor (kcal/kg min)	Weight (kg)	Total Energy Expended (kcal)

Total energy expended _____

Table 19.3 Energy Cost for Activities (Exclusive of Basal Metabolism and Specific Dynamic Action)

Type of Activity	Energy Cost (kcal/kg min)	Type of Activity	Energy Cost (kcal/kg min)
Sitting or standing still	0.010	Heavy exercise	0.070
Studying		Fast dancing	
Writing		Walking uphill	
Word processing		Jogging	
TV watching		Fast swimming	
Eating		Severe exercise	0.110
Very light activity	0.020	Tennis	
Driving car		Racquetball	
Walking slowly		Running	
Light exercise	0.025	Aerobic dancing	
Light housework		Soccer	
Walking at moderate speed		Very severe exercise	0.140
Carrying books or packages		Wrestling	
Moderate exercise	0.040	Boxing	
Fast walking		Racing	
Slow dancing		Rowing	
Slow bicycling		Full-court basketball	
Golf			

Calculating BMR/Day

The **basal metabolic rate** (**BMR**) is the rate at which kcal are spent for basal metabolism. Basal metabolism is the minimum amount of energy the body needs at rest in the fasting state. The beating of the heart, breathing, maintaining body temperature, and sending nerve impulses are some of the activities that maintain life. BMR varies according to a person's body surface area, age, and gender. On page 242, you will multiply your body surface area by a basal metabolic rate constant to arrive at your BMR/hour. This hourly rate multiplied by 24 hr/day will give you the BMR/day.

Figure 19.1 Method to estimate body surface area from height and weight.

A straight line is drawn from the subject's height (Scale 1) to the subject's weight (Scale 3). The point at which the line intersects Scale 2 is the subject's body surface area in m^2 (meters squared).

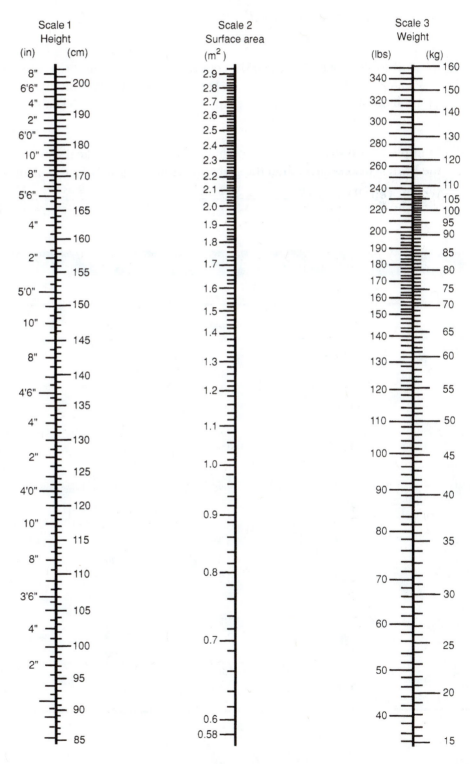

Body Surface Area

1. Use a scale to determine your weight and a measuring device (e.g., measuring tape) to determine your height. It is assumed that you are fully clothed and wearing shoes with a 1-inch heel. **Be honest about your weight.** There is no need to tell anyone else.

2. Consult Figure 19.1 and with a ruler draw a straight line from your height to your weight. The point where that line crosses the middle column shows your surface area in m^2 (squared meters). For example, a person who is 6 ft tall and weighs 170 lb has a body surface area of $1.99 m^2$.

   What is your body surface area? _____

BMR/Day

1. Consult Table 19.4 to find the BMR constant for your age and sex. Multiply your surface area by this factor to calculate your BMR/hr. For example, a 17-year-old male has a BMR constant of 41.5 kcal/m^2 hr. If his surface area is $1.99 m^2$, his BMR is $1.99 m^2 \times 41.5$ kcal/$m^2 = 82.6$ kcal/hr.

   What is your BMR/hr? _____

2. Multiply your BMR/hr by 24 to obtain the total number of kcal you need for BMR/day. For example, if the BMR is 82.6 kcal/hr, then the daily BMR rate is 24 hr $\times$ 82.6 kcal/hr = 1,982 kcal/day.

   What is your BMR/day? _____

Table 19.4 Basal Metabolic Rate Constants

Age	BMR (kcal/m^2 hr) Males	BMR (kcal/m^2 hr) Females	Age	BMR (kcal/m^2 hr) Males	BMR (kcal/m^2 hr) Females
10	47.7	44.9	29	37.7	35.0
11	46.5	43.5	30	37.6	35.0
12	45.3	42.0	31	37.4	35.0
13	44.5	40.5	32	37.2	34.9
14	43.8	39.2	33	37.1	34.9
15	42.9	38.3	34	37.0	34.9
16	42.0	37.2	35	36.9	34.8
17	41.5	36.4	36	36.8	34.7
18	40.8	35.8	37	36.7	34.6
19	40.5	35.4	38	36.7	34.5
20	39.9	35.3	39	36.6	34.4
21	39.5	35.2	40–44	36.4	34.1
22	39.2	35.2	45–49	36.2	33.8
23	39.0	35.2	50–54	35.8	33.1
24	38.7	35.1	55–59	35.1	32.8
25	38.4	35.1	60–64	34.5	32.0
26	38.2	35.0	65–69	33.5	31.6
27	38.0	35.0	70–74	32.7	31.1
28	37.8	35.0	75+	31.8	

Calculating Energy Required for SDA

The **specific dynamic action** (**SDA**) is the amount of energy needed to process food. For example, muscles that move food along the digestive tract and glands that make digestive juices use up energy. To calculate the amount of average energy you require for daily SDA, (1) add the average energy required for daily physical activity and the energy for daily BMR. (2) Multiply the total by 10% to obtain an estimated average daily SDA. For example, if 1,044 kcal/day is required for physical activity and 1,882 kcal/day is required for BMR, then an average SDA = 293 kcal/day.

What is your **average SDA/day**? _____ kcal

Calculating Average Daily Energy Requirement

Total the amounts you have calculated for average daily physical activity, daily BMR, and average daily SDA. This is your average (av.) energy requirement/day.

Av. physical activity/day _____ kcal (see p. 239)

BMR/day _____ kcal (see p. 242)

Av. SDA/day _____ kcal (see above)

Av. energy requirement/day = _____ kcal

19.3 Comparison of Average Daily Energy Intake and Average Daily Energy Requirement

Figure 19.2 illustrates that, if the average daily energy intake is the same as the average daily energy requirement, weight is likely to remain the same.

1. Compare your average daily energy intake (p. 236 or p. 238) to your average daily energy

   requirement: _____

2. Our methods of calculation are only approximate. If your two figures are within 20% of each other, you will most likely neither lose nor gain weight. If the two figures are not within 20%

   of each other, are you apt to lose weight or gain weight? _____

   Explain. _____

   _____

Figure 19.2 Comparison of average daily energy intake and average daily energy requirement.
If average daily energy intake is the same as the average daily energy requirement, the person's weight stays the same.

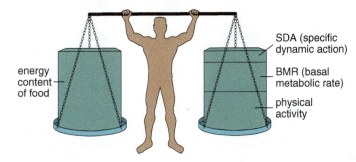

19.4   Ideal Weight

Three methods of determining your ideal weight are explained next. Your instructor will choose which one you are to use.

Ideal Weight Based on Height and Weight Tables

Tables 19.5 and 19.6 are constructed by life insurance companies from data on thousands of people, at an age when bone and muscle growth is complete. Such tables are not highly accurate. Determine your

Table 19.5 Weight and Height Table for Males				
Height		Small Frame	Medium Frame	Large Frame
(ft)	(in)	(lb)	(lb)	(lb)
5	2	128–134	131–141	138–150
5	3	130–136	133–143	140–153
5	4	132–138	135–145	142–156
5	5	134–140	137–148	144–160
5	6	136–142	139–151	146–164
5	7	138–145	142–154	149–168
5	8	140–148	145–157	152–172
5	9	142–151	148–160	155–176
5	10	144–154	151–163	158–180
5	11	146–157	154–166	161–184
6	0	149–160	157–170	164–188
6	1	152–164	160–174	168–192
6	2	155–168	164–178	172–197
6	3	158–172	167–182	176–202
6	4	162–176	171–187	181–207

Table 19.6 Weight and Height Table for Females				
Height		Small Frame	Medium Frame	Large Frame
(ft)	(in)	(lb)	(lb)	(lb)
4	10	102–111	109–121	118–131
4	11	103–113	111–123	120–134
5	0	104–115	113–126	122–137
5	1	106–118	115–129	125–140
5	2	108–121	118–132	128–143
5	3	111–124	121–135	131–147
5	4	114–127	124–138	134–151
5	5	117–130	127–141	137–155
5	6	120–133	130–144	140–159
5	7	123–136	133–147	143–163
5	8	126–139	136–150	146–167
5	9	129–142	139–153	149–170
5	10	132–145	142–156	152–173
5	11	135–148	145–159	155–176
6	0	138–151	148–162	158–179

Weight at ages 25–59 based on lowest mortality. Weight according to frame (indoor clothing weighing 3 lb is allowed for women; 5 lb for men), shoes with 1-inch heels.

ideal weight range. A person more than 10% over the weight in the tables is overweight; a person 20% over is obese. Similarly, a person 10% below the weight in the table is underweight. What is your ideal weight based on these height and weight tables?

Ideal weight _____ lb

Ideal Weight Based on Body Mass Index (BMI)

Your **body mass index** (**BMI**) is a measure of your weight relative to your height (Table 19.7). A BMI of 19–25 is considered ideal. Below 19 a person is considered underweight, and above 25 a person is considered overweight. An obese person has a BMI of 30 or greater.

Table 19.7 Body Mass Index

Height (inches)	19	20	21	22	23	24	25	26	27	28	29	30	31	32	33	34	35
				Ideal						Body Weight (pounds) Not Ideal							
58	91	96	100	105	110	115	119	124	129	134	138	143	148	153	158	162	167
59	94	99	104	109	114	119	124	128	133	138	143	148	153	158	163	168	173
60	97	102	107	112	118	123	128	133	138	143	148	153	158	163	168	174	179
61	100	106	111	116	122	127	132	137	143	148	153	158	164	169	174	180	185
62	104	109	115	120	126	131	136	142	147	153	158	164	169	175	180	186	191
63	107	113	118	124	130	135	141	146	152	158	163	169	175	180	186	191	197
64	110	116	122	128	134	140	145	151	157	163	169	174	180	186	192	197	204
65	114	120	126	132	138	144	150	156	162	168	174	180	186	192	198	204	210
66	118	124	130	136	142	148	155	161	167	173	179	186	192	198	204	210	216
67	121	127	134	140	146	153	159	166	172	178	185	191	198	204	211	217	223
68	125	131	138	144	151	158	164	171	177	184	190	197	203	210	216	223	230
69	128	135	142	149	155	162	169	176	182	189	196	203	209	216	223	230	236
70	132	139	146	153	160	167	174	181	188	195	202	209	216	222	229	236	243
71	136	143	150	157	165	172	179	186	193	200	208	215	222	229	236	243	250
72	140	147	154	162	169	177	184	191	199	206	213	221	228	235	242	250	258
73	144	151	159	166	174	182	189	197	204	212	219	227	235	242	250	257	265
74	148	155	163	171	179	186	194	202	210	218	225	233	241	249	256	264	272
75	152	160	168	176	184	192	200	208	216	224	232	240	248	256	264	272	279
76	156	164	172	180	189	197	205	213	221	230	238	246	254	263	271	279	287

Procedure to Determine Ideal Weight Based on Body Mass Index (BMI)

1. Consult Table 19.7 and find your height in inches.

2. Go across that row to your weight.

3. Read the number at the top of that column. What is your BMI? _____

4. Is your BMI within the normal range (19–25)? _____ What is your ideal weight range based on BMI? _____

Comparison

How does this weight range compare to your desirable weight range from Table 19.5 or Table 19.6?

Ideal Weight Based on Body Composition

Body composition refers to the lean body weight plus the body fat. Ideal weight based on body composition is preferred, because then you know how much of your weight is due to fat. In adult males, lean body weight should equal about 84% of weight, and body fat should be only 16% of weight. In adult females, lean body weight should equal about 77% of weight, and body fat should be only 23% of weight. Well-conditioned male athletes, such as marathon runners and swimmers, usually have 10–12% body fat, whereas football players can have as high as 19–20% body fat. At no time should the percentage of body fat drop below 5% in males and 10% in females.

To calculate **lean body weight** (**LBW**), it is necessary to subtract the amount of your body fat from your present weight. The amount of body fat is to be determined by taking skin-fold measurements, because the fat attached to the skin is roughly proportional to total body fat. The more areas of the body measured, the more accurate the estimate of body fat will be. In this laboratory, we will take only those measurements that permit you to remain fully clothed.

Procedure to Determine Lean Body Weight

1. Work in pairs. Obtain calipers designed to measure skin folds. All measurements should be done on the right side in mm (millimeters). Firmly grasp the fold of skin between the left thumb and the other four fingers and lift. Pinch and lift the fold several times to make sure no musculature is grasped. Hold the skin fold firmly, and place the contact side of the calipers $1/2$ inch below the thumb and fingers; do not let go of the fold. Close the calipers on the skin fold, and do the measurements noted in items 2–6 following. Take the reading to the nearest $1/2$ mm. Release the grip on the caliper and release the fold. To make sure the reading is accurate, repeat the measurement a few times. If the second measurement is within 1–2 mm of the first, it is reliable. Record two readings.
2. **Triceps (females only).** The triceps skin fold is measured on the back of the upper arm, halfway between the elbow and the tip of the shoulder, while the arm is hanging loosely at the subject's side. Grasp the skin fold parallel to the long axis of the arm and measure as described in step 1. Record in #7 below. Record two readings.
3. **Ilium—hip (males and females).** Fold the skin diagonally just above the top of the hip bone on an imaginary line that would divide the body into front and back halves. Measure and record as before. Record two readings.
4. **Abdomen (males and females).** Fold the skin vertically 1 inch to the right of the navel. Measure and record as before. Record two readings.
5. **Chest (males only).** Fold the skin diagonally, midway between the nipple and the armpit. Measure and record as before. Record two readings.
6. **Axilla—side (males only).** Fold the skin vertically at the level of nipple on an imaginary line that would divide the body into front and back halves. Measure and record as before. Record two readings.
7. Total any four of your skin-fold measurements.

	Females		**Males**	
Triceps	_____ _____ mm			
Ilium	_____ _____ mm		_____ _____ mm	
Abdomen	_____ _____ mm		_____ _____ mm	
Chest			_____ _____ mm	
Axilla			_____ _____ mm	
Total of any four skin-fold measurements	_____ mm		_____ mm	

8. Consult Table 19.8 (males) and Table 19.9 (females) to determine percentage (%) body fat from the sum of four skin-fold measurements and your age.

What is your % body fat? _____ %

Table 19.8 Percentage Fat Estimates for Males

Sum of Four Skin Folds (mm)	Age								
	18–22	23–27	28–32	33–37	38–42	43–47	48–52	53–57	58–older
8–12	1.9%	2.5%	3.2%	3.8%	4.4%	5.0%	5.7%	6.3%	6.9%
13–17	3.3%	3.9%	4.5%	5.1%	5.7%	6.4%	7.0%	7.6%	8.2%
18–22	4.5%	5.2%	5.8%	6.4%	7.0%	7.7%	8.3%	8.9%	9.5%
23–27	5.8%	6.4%	7.1%	7.7%	8.3%	8.9%	9.5%	10.2%	10.8%
28–32	7.1%	7.7%	8.3%	8.9%	9.5%	10.2%	10.8%	11.4%	12.0%
33–37	8.3%	8.9%	9.5%	10.1%	10.8%	11.4%	12.0%	12.6%	13.2%
38–42	9.5%	10.1%	10.7%	11.3%	11.9%	12.6%	13.2%	13.8%	14.4%
43–47	10.6%	11.6%	11.9%	12.5%	13.1%	13.7%	14.4%	15.0%	15.6%
48–52	11.8%	12.4%	13.0%	13.6%	14.2%	14.9%	15.5%	16.1%	16.7%
53–57	12.9%	13.5%	14.1%	14.7%	15.4%	16.0%	16.6%	17.2%	17.9%
58–62	14.0%	14.6%	15.2%	15.8%	16.4%	17.1%	17.7%	18.3%	18.9%
63–67	15.0%	15.6%	16.3%	16.9%	17.5%	18.1%	18.8%	19.4%	20.0%
68–72	16.1%	16.7%	17.3%	17.9%	18.5%	19.2%	19.8%	20.4%	21.0%
73–77	17.1%	17.7%	18.3%	18.9%	19.5%	20.2%	20.8%	21.4%	22.0%
78–82	18.0%	18.7%	19.3%	19.9%	20.5%	21.0%	21.8%	22.4%	23.0%
83–87	19.0%	19.6%	20.2%	20.8%	21.5%	22.1%	22.7%	23.3%	24.0%
88–92	19.9%	20.5%	21.2%	21.8%	22.4%	23.0%	23.6%	24.3%	24.9%
93–97	20.8%	21.4%	22.1%	22.7%	23.3%	23.9%	24.8%	25.2%	25.8%
98–102	21.7%	22.6%	22.9%	23.5%	24.2%	24.8%	25.4%	26.0%	26.7%
103–107	22.5%	23.2%	23.8%	24.4%	25.0%	25.6%	26.3%	26.9%	27.5%
108–112	23.4%	24.0%	24.6%	25.2%	25.8%	26.5%	27.1%	27.7%	28.3%
113–117	24.1%	24.8%	25.4%	26.0%	26.6%	27.3%	27.9%	28.5%	29.1%
118–122	24.9%	25.5%	26.2%	26.8%	27.4%	28.0%	28.6%	29.3%	29.9%
123–127	25.7%	26.3%	26.9%	27.5%	28.1%	28.8%	29.4%	30.0%	30.6%
128–132	26.4%	27.0%	27.6%	28.2%	28.8%	29.5%	30.1%	30.7%	31.3%
133–137	27.1%	27.7%	28.3%	28.9%	29.5%	30.2%	30.8%	31.4%	32.0%
138–142	27.7%	28.3%	29.0%	29.6%	30.2%	30.8%	31.4%	32.1%	32.7%
143–147	28.3%	29.0%	29.6%	30.2%	30.8%	31.5%	32.1%	32.7%	33.3%
148–152	29.0%	29.6%	30.2%	30.8%	31.4%	32.7%	32.7%	33.3%	33.9%
153–157	29.5%	30.2%	30.8%	31.4%	32.0%	32.7%	33.3%	33.9%	34.5%
158–162	30.1%	30.7%	31.3%	31.9%	32.6%	33.2%	33.8%	34.4%	35.1%
163–167	30.6%	31.2%	31.9%	32.5%	33.1%	33.7%	34.3%	35.0%	35.6%
168–172	31.1%	31.7%	32.4%	33.0%	33.6%	34.2%	34.8%	35.5%	36.1%
173–177	31.6%	32.2%	32.8%	33.5%	34.1%	34.7%	35.3%	35.9%	36.6%
178–182	32.0%	32.7%	33.3%	33.9%	34.5%	35.2%	35.8%	36.4%	37.0%
183–187	32.5%	33.1%	33.7%	34.3%	34.9%	35.6%	36.2%	36.8%	37.4%
188–192	32.9%	33.5%	34.1%	34.7%	35.3%	36.0%	36.6%	37.2%	37.8%
193–197	33.2%	33.8%	34.5%	35.1%	35.7%	36.3%	37.0%	37.8%	38.2%
198–202	33.6%	34.2%	34.8%	35.4%	36.1%	36.7%	37.3%	37.9%	38.5%
203–207	33.9%	34.5%	35.1%	35.7%	36.4%	37.0%	37.6%	38.2%	38.9%

Table 19.9　Percentage Fat Estimates for Females

Sum of Four Skin Folds (mm)	Age								
	18–22	23–27	28–32	33–37	38–42	43–47	48–52	53–57	58–older
8–12	8.8%	9.0%	9.2%	9.4%	9.5%	9.7%	9.9%	10.1%	10.3%
13–17	10.8%	10.9%	11.1%	11.3%	11.5%	11.7%	11.8%	12.0%	12.2%
18–22	12.6%	12.8%	13.0%	13.2%	13.4%	13.5%	13.7%	13.9%	14.1%
23–27	14.5%	14.6%	14.8%	15.0%	15.2%	15.4%	15.6%	15.7%	15.9%
28–32	16.2%	16.4%	16.6%	16.8%	17.0%	17.1%	17.3%	17.5%	17.7%
33–37	17.9%	18.1%	18.3%	18.5%	18.7%	18.9%	19.0%	19.2%	19.4%
38–42	19.6%	19.8%	20.0%	20.2%	20.3%	20.5%	20.7%	20.9%	21.1%
43–47	21.2%	21.4%	21.6%	21.8%	21.9%	22.1%	22.3%	22.5%	22.7%
48–52	22.8%	22.9%	23.1%	23.3%	23.5%	23.7%	23.8%	24.0%	24.2%
53–57	24.2%	24.4%	24.6%	24.8%	25.0%	25.2%	25.3%	25.5%	25.7%
58–62	25.7%	25.9%	26.0%	26.2%	26.4%	26.6%	26.8%	27.0%	27.1%
63–67	27.1%	27.2%	27.4%	27.6%	27.8%	28.0%	28.2%	28.3%	28.5%
68–72	28.4%	28.6%	28.7%	28.9%	29.1%	29.3%	29.5%	29.7%	29.8%
73–77	29.6%	29.8%	30.0%	30.2%	30.4%	30.6%	30.7%	30.9%	31.1%
78–82	30.9%	31.0%	31.2%	31.4%	31.6%	31.8%	31.9%	32.1%	32.3%
83–87	32.0%	32.2%	32.4%	32.6%	32.7%	32.9%	33.1%	33.3%	33.5%
88–92	33.1%	33.3%	33.5%	33.7%	33.8%	34.0%	34.2%	34.4%	34.6%
93–97	34.1%	34.3%	34.5%	34.7%	34.9%	35.1%	35.2%	35.4%	35.6%
98–102	35.1%	35.3%	35.5%	35.7%	35.9%	36.0%	36.2%	36.4%	36.6%
103–107	36.1%	36.2%	36.4%	36.6%	36.8%	37.0%	37.2%	37.3%	37.5%
108–112	36.9%	37.1%	37.3%	37.5%	37.7%	37.9%	38.0%	38.2%	38.4%
113–117	37.8%	37.9%	38.1%	38.3%	39.2%	39.4%	39.6%	39.8%	40.0%
118–122	38.5%	38.7%	38.9%	39.1%	39.4%	39.6%	39.8%	40.0%	40.5%
123–127	39.2%	39.4%	39.6%	39.8%	40.0%	40.1%	40.3%	40.5%	40.7%
128–132	39.9%	40.1%	40.2%	40.4%	40.6%	40.8%	41.0%	41.2%	41.3%
133–137	40.5%	40.7%	40.8%	41.0%	41.2%	41.4%	41.6%	41.7%	41.9%
138–142	41.0%	41.2%	41.4%	41.6%	41.7%	41.9%	42.1%	42.3%	42.5%
143–147	41.5%	41.7%	41.9%	42.0%	42.2%	42.4%	42.6%	42.8%	43.0%
148–152	41.9%	42.1%	42.3%	42.8%	42.6%	42.8%	43.0%	43.2%	43.4%
153–157	42.3%	42.5%	42.6%	42.8%	43.0%	43.2%	43.4%	43.6%	43.7%
158–162	42.6%	42.8%	43.0%	43.1%	43.3%	43.5%	43.7%	43.9%	44.1%
163–167	42.9%	43.0%	43.2%	43.4%	43.6%	43.8%	44.0%	44.1%	44.3%
168–172	43.1%	43.2%	43.4%	43.6%	43.8%	44.0%	44.2%	44.3%	44.5%
173–177	43.2%	43.4%	43.6%	43.8%	43.9%	44.1%	44.3%	44.5%	44.7%
178–182	43.3%	43.5%	43.7%	43.8%	44.0%	44.2%	44.4%	44.6%	44.8%

9. Multiply your present body weight in pounds (lb) by your present % body fat to determine how much of your body weight is fat. For example, if a male weighs 180 lb and his percentage body fat is 20%, then 180 × .20 = 36 lb. His body fat is 36 lb.

How much of your body weight is fat? _____ lb

10. Subtract the weight of fat from your present body weight. This is lean body weight (LBW). For example, if a male weighs 180 lb and his body fat is 36 lb, then 180 − 36 = 144. His LBW is 144 lb.

What is your **LBW?** _____ lb

11. Calculate ideal weight based on body composition.

For males, 84% of ideal weight should be LBW. Therefore, ideal weight = LBW/0.84. What is your ideal weight based on body composition? For example, if a male has a LBW of 144 lb, his ideal weight is 144/.84 = 171 lb.

> Male ideal weight = _____ lb

(At this weight, your body will contain 16% fat, the amount generally recommended for males.)

For females, 77% of ideal weight should be LBW. Therefore, ideal weight = LBW/0.77. What is your ideal weight based on body composition?

> Female ideal weight = _____ lb

(At this weight, your body will contain 23% fat, the amount generally recommended for females.)

Comparison

How does your ideal weight based on body composition compare to your ideal weight based on BMI?

_____

_____

Conclusions: Ideal Body Weight

- What is your average daily energy requirement? _____ kcal (p. 243)
- What is your average daily energy intake? _____ kcal (p. 236)
- What is your ideal weight? (Take an average of any you have calculated.) _____ lb

  For your information, 1 lb of fat represents 3,500 kcal. Therefore, if you want to lose 1 lb of fat, you must either reduce your intake by 3,500 kcal or increase your activity by 3,500 kcal. Assuming the same amount of physical activity, you could lose 1 lb per week by reducing your kcal intake by 500 kcal per day or gain 1 lb per week by increasing your kcal intake by 500 kcal per day.

- What is your recommendation for maintaining or achieving your ideal weight? _____

_____

Laboratory Review 19

1. What is a listing of all foods eaten for a day? _____

2. What does *kcal* mean? _____

3. Daily energy requirement includes energy for physical activity, SDA, and what else? _____

4. Give an example of a basal metabolism activity. _____

5. *SDA* refers to energy needed for what activity? _____

6. With age, the BMR decreases. What is the implication for daily energy intake? _____

7. Why does a tall, thin person have a higher BMR than a short, stout person? _____

8. Daily energy requirement must be in balance with what to not gain weight? _____

9. If the average energy intake/day equals the average energy requirement/day, the

    person will not _____.

10. Ideal weight includes lean body weight and what else? _____

11. Generally speaking, a male should have no more than what percentage body fat? _____

12. Calculation of ideal weight is considered to be the most accurate if it is based on your _____.

13. How many kcal are equal to a pound of fat? _____

14. In which sex is it natural to have a higher percentage of body fat? _____

15. Why do males typically have a higher basal metabolic rate than females? _____

16. Why is energy needed when the body is at rest? _____

17. Why is energy needed to process food? _____

18. Why should someone monitor his or her average daily energy intake when attempting to maintain average

    body weight? _____

Essentials of Biology Website

Instructors can find lab prep information and answers to all of the laboratory questions in the Laboratory Resource Guide. *Students* can practice their knowledge with quizzes, animations, flashcards, and much more.

www.mhhe.com/maderessentials4

McGraw-Hill Access Science Website

An online encyclopedia of science and technology that provides information, including videos, that can enhance the laboratory experience.

www.accessscience.com

20

Basic Mammalian Anatomy II

Learning Outcomes

20.1 Cardiovascular System
- Locate and identify the chambers of the heart and the vessels connected to these chambers.
- Name and locate the valves of the heart, and trace the path of blood through the heart.
- Name the major vessels of the pulmonary and systemic circuits of the cardiovascular system, and trace the path of blood to various organs.
- Compare the pulmonary and systemic circuits of vertebrates.

20.2 Respiratory, Digestive, and Urinary Systems
- Locate and identify in a fetal pig the organs of the respiratory system, and state a function for each organ.
- Locate and identify the organs of the digestive system, and state a function for each organ.
- Locate and identify the organs of the urinary system, and state a function for each organ.

20.3 Male Reproductive System
- Locate and identify the organs of the male reproductive system in a fetal pig.
- State a function for the organs of the male reproductive system.
- Compare the pig reproductive system with that of the human male.

20.4 Female Reproductive System
- Locate and identify the organs of the female reproductive system in a fetal pig.
- State a function for the organs of the female reproductive system.
- Compare the pig reproductive system with that of a human female.

Introduction

The heart and blood vessels form the cardiovascular system. The heart is a double pump that keeps the blood flowing in one direction—away from and then back to the heart. The blood vessels transport blood and its contents, serve the needs of the body's cells by carrying out exchanges with them, and direct blood flow to those systemic tissues that most require it at the moment. After removing the heart and examining the heart chambers, you will have an opportunity to remove and study the lungs and the small intestine. The latter will expose the urinary and reproductive systems. These two systems are so closely associated in mammals that they are often considered together as the urogenital system. In this laboratory, we will focus first on dissecting the urinary and reproductive systems in the fetal pig. We will then compare the anatomy of the reproductive systems in pigs with those in humans.

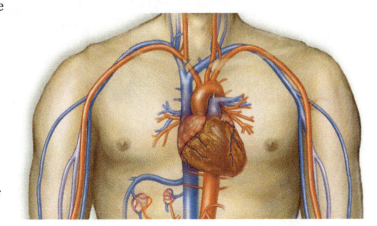

20.1   Cardiovascular System

The **cardiovascular system** includes the heart and two major circular pathways: the **pulmonary circuit** to and from the lungs and the **systemic circuit** to and from the body's organs.

Heart

The mammalian heart has a **right** and **left atrium** and a **right** and **left ventricle.** *To tell the left from the right side, mentally position the heart so that it corresponds to your own body.* Contraction of the heart pumps the blood through the heart and out into the arteries. The right side of the heart sends blood through the smaller pulmonary circuit, and the left side of the heart sends blood through the much larger systemic circuit.

Observation: Heart

Heart Model

1. Study a heart model (Fig. 20.1) or a preserved heart, and identify the four chambers of the heart: right atrium, right ventricle, left atrium, left ventricle.

2. Which ventricle is more muscular? _____ Why is this appropriate? _____

_____

Figure 20.1   External view of the mammalian heart.
Externally, notice the coronary arteries and cardiac veins that serve the heart itself.

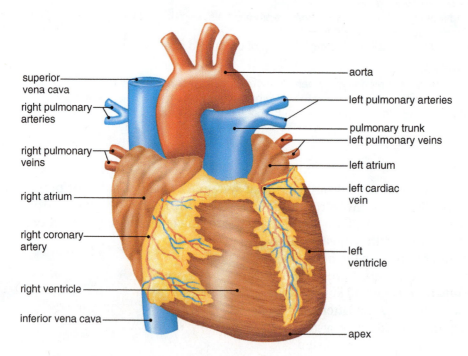

superior vena cava
right pulmonary arteries
right pulmonary veins
right atrium
right coronary artery
right ventricle
inferior vena cava

aorta
left pulmonary arteries
pulmonary trunk
left pulmonary veins
left atrium
left cardiac vein
left ventricle
apex

3. Identify the vessels connected to the heart. Locate the following structures:
   a. **Pulmonary trunk:** Leaves the ventral side of the heart from the top of the right ventricle and then passes forward diagonally before branching into the right and left pulmonary arteries.
   b. **Aorta:** Arises from the anterior end of the left ventricle, just dorsal to the origin of the pulmonary trunk. The aorta soon bends to the animal's left as the **aortic arch.**
   c. **Venae cavae:** Anterior (superior) and posterior (inferior) venae cavae enter the right atrium. They take blood from the head and body, respectively, to the heart.
   d. **Pulmonary veins:** Return blood from the lungs to the left atrium. Why are the pulmonary veins colored red in Figure 20.2? _____

   _____

4. Identify the four chambers of the heart in longitudinal section: right atrium, right ventricle, left atrium, and left ventricle, and *label them in Figure 20.2.*

5. Locate the valves of the heart (Fig. 20.2). Remove the ventral half of the heart model or the ventral half of the heart. Identify the following structures:
   a. **Right atrioventricular (tricuspid) valve** and **left atrioventricular (bicuspid, mitral) valve**
   b. **Pulmonary semilunar valve** (in the base of the pulmonary trunk)
   c. **Aortic semilunar valve** (in the base of the aorta)
   d. **Chordae tendineae:** Hold the atrioventricular valves in place while the heart contracts. These extend from the papillary muscles.

Figure 20.2   Internal view of the mammalian heart.
Internally, the heart has four chambers, and a septum separates the left side from the right side.

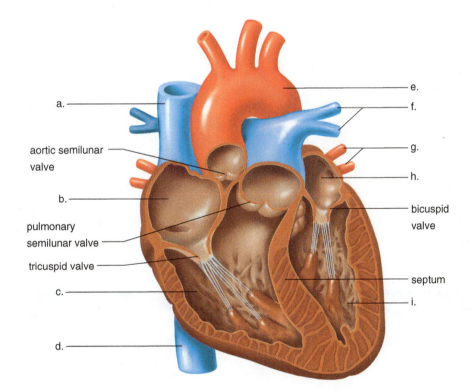

Tracing the Path of Blood Through the Heart

To demonstrate that O_2-poor blood is kept separate from O_2-rich blood, trace the path of blood from the right side of the heart to the aorta by filling in the blanks with the names of blood vessels and valves. In the adult mammal, note that blood passes through the lungs to go from the right side of the heart to the left side.

From Venae Cavae

_____

_____ **valve**

_____

_____ **valve**

_____

_____

To Lungs

From Lungs

_____

_____

_____ **valve**

_____

_____ **valve**

To Aorta

Pulmonary Circuit and Systemic Circuit

Blood that leaves the heart enters one of two sets of blood vessels: the **pulmonary circuit,** which takes blood from the heart to the lungs and from the lungs to the heart, and the **systemic circuit,** which takes blood from the heart to the body proper and from the body proper to the heart.

Pulmonary Circuit

1. Sequence the blood vessels in the pulmonary circuit to trace the path of blood from the right ventricle to the left atrium of the heart.

   Right ventricle of the heart

   _____

   _____

   Lungs

   _____

   Left atrium of the heart

2. Which of these blood vessels contains O_2-rich blood? _____

Systemic Circuit

1. The major blood vessels in the systemic system are the **aorta** (takes blood away from the heart) and the **venae cavae** (take blood to the heart). Otherwise, with the help of Figure 20.3, complete Table 20.1.

Table 20.1	Other Blood Vessels in the Systemic Circuit	
Body Part	**Artery**	**Vein**
Head		
Front legs in pig (arms in humans)		
Kidney		
Hind legs		

Figure 20.3 Overview of the arteries and veins in mammals.
Arteries and veins are found in all parts of the body. In this drawing, only veins are on the left, and only arteries are on the right.

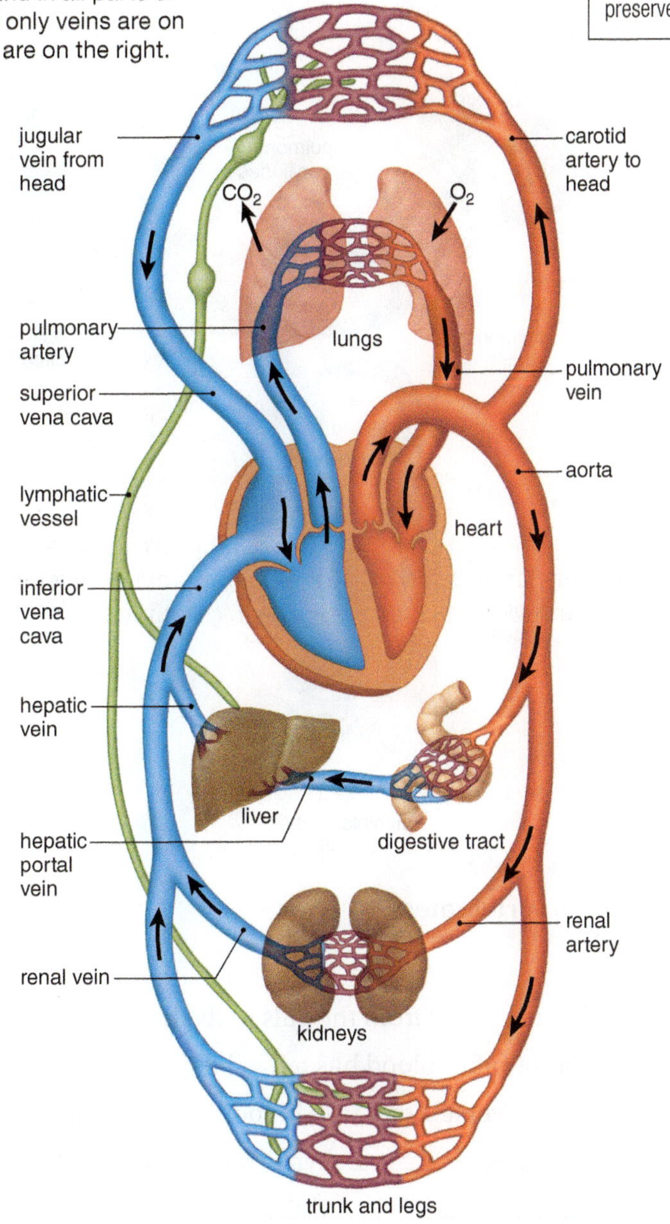

jugular vein from head
CO_2
O_2
carotid artery to head
pulmonary artery
lungs
pulmonary vein
superior vena cava
aorta
lymphatic vessel
heart
inferior vena cava
hepatic vein
liver
hepatic portal vein
digestive tract
renal artery
renal vein
kidneys
trunk and legs

> **Latex gloves** Wear goggles, latex gloves, and protective clothing when handling preserved animal organs.

2. With the help of Figure 20.3 and Table 20.1, sequence the blood vessels in the systemic circuit to trace the path of blood from the heart to the kidneys and from the kidneys to the heart.

   Left ventricle of the heart

   ____

   ____

   Kidneys

   ____

   ____

   Right atrium of the heart

Vertebrate Cardiovascular Systems

Compare the cardiovascular systems of the vertebrates in Figure 20.4 and answer these questions.

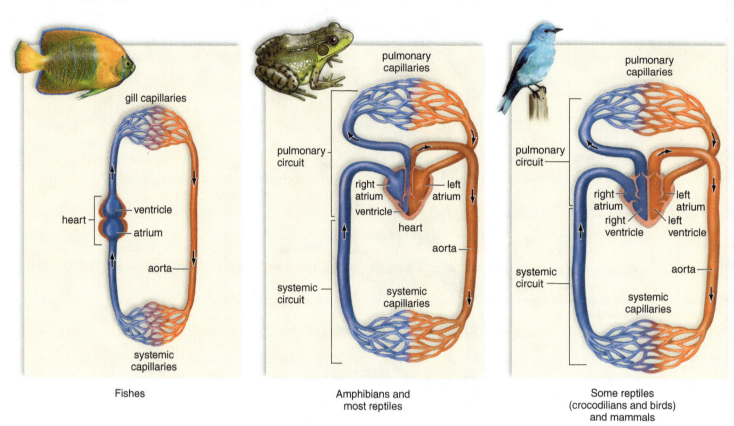

Figure 20.4 Cardiovascular systems in vertebrates.

1. Do fish have a blood vessel that returns blood from the gills to the heart? _____ Would you expect blood pressure to be high or low after blood has moved through the gills? _____

2. What animals studied have pulmonary vessels that take blood from the heart to the respiratory organ and back to the heart? _____ What is the advantage of a pulmonary circuit? _____

3. Which of these animals has a four-chambered heart? _____ What is the advantage of having separate ventricles? _____ _____

4. The circulatory system distributes the heat of muscle contraction in birds and mammals. Is the anatomy of birds and mammals conducive to maintaining a warm internal temperature? _____ Explain your answer. _____ _____

20.2 Respiratory, Digestive, and Urinary Systems

You will now have the opportunity to examine more carefully the organs of the respiratory and digestive systems you merely observed in Laboratory 17. You will also study the urinary system.

> **Observation: Organs of the Respiratory System and Digestive System**

Respiratory System

1. Open a pig's mouth, insert your blunt probe into the **glottis,** and explore the pathway of air in normal breathing by carefully working the probe down through the **larynx** to the level of the bronchi.
2. If you wish, you can slit open the larynx along its midline and observe the small, paired, lateral flaps inside, known as **vocal cords.** These are not yet well developed in this fetal animal.
3. Remove the heart of your pig by cutting the blood vessels that enter and exit heart. Name these blood vessels as you cut them. You will now see the trachea and how it divides into two bronchi. Carefully cut the trachea crosswise just below the larynx. Do not cut the esophagus, which lies just dorsal to the trachea.
4. Lift out the entire portion of the respiratory system you have just freed: trachea, bronchi, bronchioles, and lungs. Place them in a small container of water. Holding the trachea with your forceps, gently but firmly stroke the lung repeatedly with the blunt wooden base of one of your probes. If you work carefully, the alveolar tissue will be fragmented and rubbed away, leaving the branching system of air tubes and blood vessels.

Digestive System

1. With removal of the heart and lungs, you now have a better view of the esophagus. Open the mouth again, and insert a blunt probe into the **esophagus** (see Fig. 17.3). Then trace the esophagus to the stomach.
2. Open one side of the **stomach,** and examine its interior surface. Does it appear smooth or rough?

   _____

3. At the lower end of the stomach, find the **pyloric sphincter,** the muscle that surrounds the entrance to the upper region of the small intestine, the **duodenum.** The pyloric sphincter regulates the entrance of material into the duodenum from the stomach.
4. Find the gallbladder, which is a small, greenish sac dorsal to the liver you also observed in Laboratory 17, page 220. Dissect more carefully now to find the **bile duct system** (see Fig. 17.5), which conducts the bile to the duodenum.
5. Locate the rest of the **small intestine** and the **large intestine** (colon and rectum), first observed in Laboratory 17, page 222. To complete your study, sever the coiled small intestine just below the duodenum, and sever the colon at the point where it joins the **rectum.**
6. Carefully cut the mesenteries holding the coils of the small intestine so that it can be laid out in a straight line. As defined previously, mesenteries are double-layered sheets of membrane that project from the body wall and support the organs.
7. Measure and record in meters the length of the intestinal tract. _____ m
8. Considering that nutrient molecules are absorbed into the blood vessels of the small intestine, why would such a great length be beneficial to the body? _____
9. Slit open a short portion of the intestines, and note the corrugated texture of the interior lining.

Figure 20.5 Urinary system of the fetal pig.

In both sexes, urine is made by the kidneys, transported to the bladder by the ureters, stored in the bladder, and then excreted from the body through the urethra.

Female

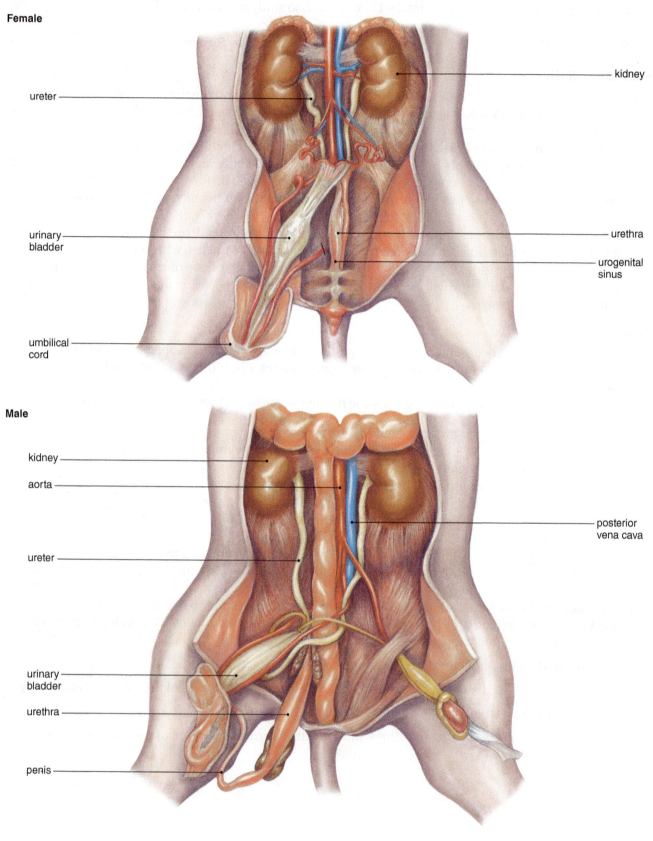

kidney

ureter

urinary
bladder

urethra

urogenital
sinus

umbilical
cord

Male

kidney

aorta

ureter

posterior
vena cava

urinary
bladder

urethra

penis

Urinary System in Pigs

The urinary system consists of the **kidneys,** which produce urine; the **ureters,** which transport urine to the **urinary bladder,** where urine is stored; and the **urethra,** which transports urine to the outside. In males, the urethra also transports sperm during copulation.

Observation: Urinary System in Pigs

During this dissection, compare the urinary system structures of male and female fetal pigs. Later in this laboratory period, exchange specimens with a neighboring team for a more thorough inspection.

1. The large, paired kidneys (Fig. 20.5) are reddish organs covered by **peritoneum,** a membrane that anchors them to the dorsal wall of the abdominal cavity, sometimes called the **peritoneal cavity.** Clean the peritoneum away from one of the kidneys, and study it more closely.
2. Locate the **ureters,** which leave the kidneys and run posteriorly under the peritoneum (Fig. 20.5).
3. Clear the peritoneum away, and follow a ureter to the **urinary bladder,** which normally lies in the ventral portion of the abdominal cavity. The urinary bladder is on the inner surface of the flap of tissue to which the umbilical cord was attached.
4. The **urethra** arises from the bladder posteriorly and joins the urogenital sinus. Follow the urethra until it passes from view into the ring formed by the pelvic girdle.
5. Sequence the organs in the urinary system to trace the path of urine from its production to its exit.

---

6. Using a scalpel, section one of the kidneys in place, cutting it lengthwise (Fig. 20.6). At the center of the medial portion of the kidney is an irregular, cavity-like reservoir, the **renal pelvis.** The outermost portion of the kidney (the **renal cortex**) shows many small striations perpendicular to the outer surface. The cortex and the more even-textured **renal medulla** contain **nephrons** (excretory tubules), microscopic organs that produce urine.

Figure 20.6   Anatomy of the kidney.
A kidney has a renal cortex, a renal medulla, a renal pelvis, and microscopic tubules called nephrons.

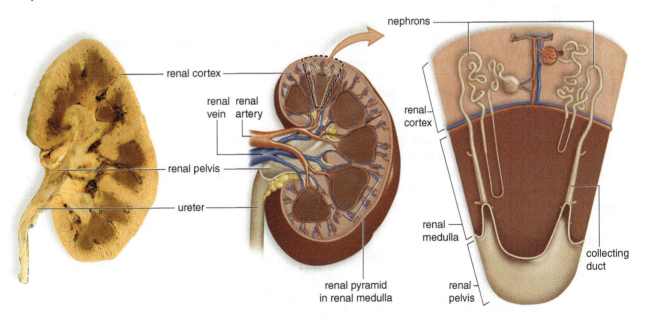

20.3 Male Reproductive System

The **male reproductive system** consists of the **testes** (sing., **testis**), which produce sperm, and the **epididymides** (sing., **epididymis**), which store sperm before they enter the **vasa deferentia** (sing., **vas deferens**). Just prior to ejaculation, sperm leave the vasa deferentia and enter the **urethra,** located in the penis. The **penis** is the male organ of sexual intercourse. **Seminal vesicles,** the **prostate gland,** and the **bulbourethral glands** (Cowper's glands) add fluid to semen (sperm plus fluids) after sperm reach the urethra. Table 20.2 summarizes the male reproductive organs.

The testes begin their development in the abdominal cavity, just anterior and dorsal to the kidneys. Before birth, however, they gradually descend into paired **scrotal sacs** within the scrotum, suspended anterior to the anus. Each scrotal sac is connected to the body cavity by an **inguinal canal,** the opening of which can be found in your pig. The passage of the testes from the body cavity into the scrotal sacs is called the descent of the testes, and it occurs in human males. The testes in most of the male fetal pigs being dissected will probably be partially or fully descended.

Observation: Male Reproductive System in Pigs

While doing this dissection, consult Table 20.2 for the function of the male reproductive organs.

Inguinal Canal, Testis, Epididymis, and Vas Deferens

1. Locate the opening of the left inguinal canal, which leads to the left scrotal sac (Fig. 20.7).
2. Expose the canal and sac by making an incision through the skin and muscle layers from a point over this opening back to the left scrotal sac.
3. Open the sac, and find the testis. Note the much-coiled tubule—the epididymis—that lies alongside a testis. An epididymis is continuous with a vas deferens, which runs toward the abdominal cavity.
4. Each vas deferens loops over an umbilical artery and ureter and unites with the urethra as it leaves the urinary bladder. We will dissect this juncture below.

Table 20.2	Male Reproductive Organs and Functions
Organ	**Function**
Testis	Produces sperm and sex hormones
Epididymis	Stores sperm as they mature
Vas deferens	Stores sperm and conducts sperm to urethra
Urethra	Conducts sperm to urogenital opening
Seminal vesicle Prostate gland Bulbourethral glands	Contributes secretions to semen
Penis	Organ of copulation

Penis, Urethra, and Accessory Glands

1. Cut through the ventral skin surface just posterior to the umbilical cord. This will expose the rather undeveloped penis, which contains a long portion of the urethra.
2. Lay the penis to one side, and then cut down through the ventral midline, laying the legs wide apart in the process (Fig. 20.8). The cut will pass between muscles and through pelvic cartilage (bone has not developed yet). Do not cut any of the ducts or tracts in the region.
3. You will now see the urethra ventral to the rectum. It is somewhat heavier in the male due to certain accessory glands:
   a. Bulbourethral glands (Cowper's glands), about 1 cm in diameter, are further along the urethra and are more prominent than the other accessory glands.
   b. The prostate gland, about 4 mm across and 3 mm thick, is located on the dorsal surface of the urethra, just posterior to the juncture of the bladder with the urethra. It is often difficult to locate and is not shown in Figures 20.7 and 20.8.
   c. Small, paired seminal vesicles may be seen on either side of the prostate gland.

Figure 20.7 Male reproductive system of the fetal pig.

In males, the urinary system and the reproductive system are joined. The vasa deferentia (sing., vas deferens) enter the urethra, which also carries urine.

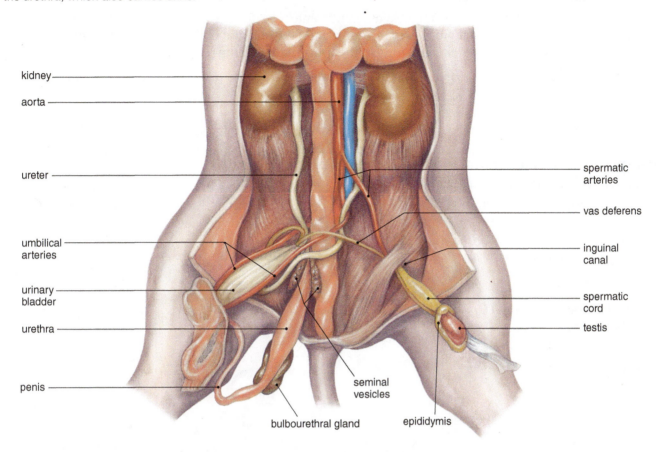

kidney
aorta
ureter
umbilical arteries
urinary bladder
urethra
penis
bulbourethral gland
seminal vesicles
epididymis
spermatic arteries
vas deferens
inguinal canal
spermatic cord
testis

Figure 20.8 Photograph of the male reproductive system of the fetal pig.

Compare the diagram in Figure 20.7 to this photograph to help identify the structures of the male urogenital system.

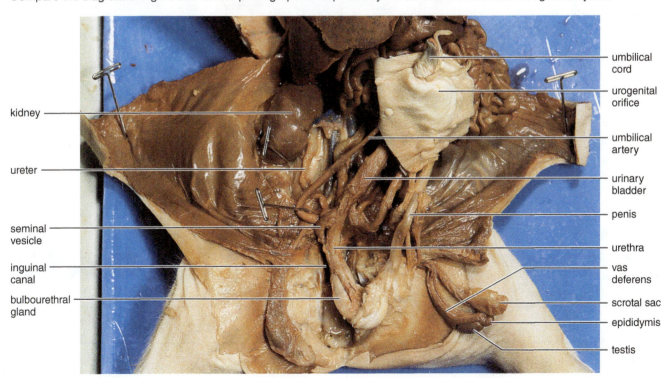

kidney
ureter
seminal vesicle
inguinal canal
bulbourethral gland
umbilical cord
urogenital orifice
umbilical artery
urinary bladder
penis
urethra
vas deferens
scrotal sac
epididymis
testis

4. Trace the urethra as it leaves the bladder. When it nears the end of the abdominal cavity, it turns rather abruptly and runs anteriorly just under the skin where you have just dissected it. This portion of the urethra is within the penis.

5. Now you should be able to see the vasa deferentia enter the urethra. If necessary, free these ducts from surrounding tissue to see them enter the urethra near the location of the prostate gland. In males, the urethra transports both sperm and urine to the urogenital opening (Fig. 17.1).

6. Sequence the organs in the male reproductive system to trace the path of sperm from the organ of production to the penis. _____

_____

Comparison of Male Fetal Pig and Human Male

Use Figure 20.9 to help you compare the male pig reproductive system with the human male reproductive system. Complete Table 20.3, which compares the location of the penis in these two mammals.

Figure 20.9   Human male urogenital system.
In the fetal pig, but not in the human male, the penis lies beneath the skin and exits at a urogenital opening.

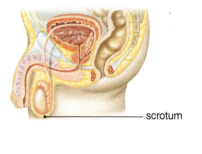

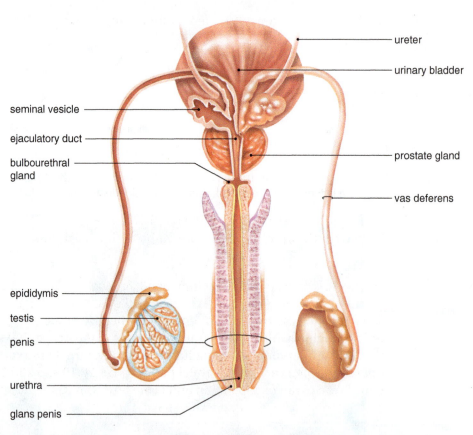

Table 20.3 Location of Penis in Male Fetal Pig and Human Male	
Fetal Pig	**Human**
Penis	

20.4 Female Reproductive System

The **female reproductive system** (Table 20.4) consists of the **ovaries,** which produce eggs, and the **oviducts,** which transport eggs to the **uterus,** where development occurs. In the fetal pig, the uterus does not form a single organ, as in humans, but is partially divided into external structures called **uterine horns,** which connect with the oviduct. The **vagina** is the birth canal and the female organ of sexual intercourse.

Table 20.4 Female Reproductive Organs and Functions

Organ	Function
Ovary	Produces egg and sex hormones
Oviduct (fallopian tube)	Conducts egg toward uterus
Uterus	Houses developing offspring
Vagina	Receives penis during copulation and serves as birth canal

Observation: Female Reproductive System in Pigs

While doing this dissection, consult Table 20.4 for the function of the female reproductive organs.

Ovaries and Oviducts

1. Locate the paired ovaries, small bodies suspended from the peritoneal wall in mesenteries, posterior to the kidneys (Figs. 20.10 and 20.11).
2. Closely examine one ovary. Note the small, short, coiled **oviduct,** sometimes called the fallopian tube. The oviduct does not attach directly to the ovary but ends in a funnel-shaped structure with fingerlike processes (fimbriae) that partially encloses the ovary. The egg produced by an ovary enters an oviduct, where it is fertilized by a sperm, if reproduction will occur. Any resulting embryo passes to the uterus.

Uterine Horns

1. Locate the **uterine horns.** (Do not confuse the uterine horns with the oviducts; the latter are much smaller and are found very close to the ovaries.)
2. Find the body of the uterus located where the uterine horns join.

Vagina

1. Separate the hindlimbs of your specimen, and cut down along the midventral line. The cut will pass through muscle and the cartilaginous pelvic girdle. With your fingers, spread the cut edges apart, and use blunt dissecting instruments to separate connective tissue.
2. Now find the vagina, which passes from the uterus to the **urogenital sinus.** The vagina is dorsal to the urethra, which also enters the urogenital sinus, and ventral to the **rectum,** which exits at the **anus.** The urogenital sinus opens at the urogenital papilla (Figs. 20.10 and 20.11).
3. The vagina is the organ of copulation and is the birth canal. The receptacle for the vagina in a pig, the urogenital sinus, is absent in adult humans and several other adult female mammals in which both the urethra and the vagina have their own openings.
4. The vagina plays a critical role in reproduction, even though development of the offspring occurs in the uterus. Explain. _____

_____

Figure 20.10 Female reproductive system of the fetal pig.

In a pig, both the vagina and the urethra enter the urogenital sinus, which opens at the urogenital papilla.

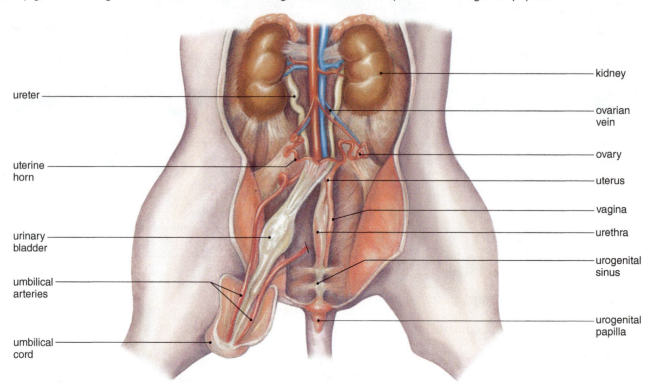

ureter

uterine horn

urinary bladder

umbilical arteries

umbilical cord

kidney

ovarian vein

ovary

uterus

vagina

urethra

urogenital sinus

urogenital papilla

Figure 20.11 Photograph of the female reproductive system of the fetal pig.

Compare the diagram in Figure 20.10 with this photograph to help identify the structures of the female urinary and reproductive systems.

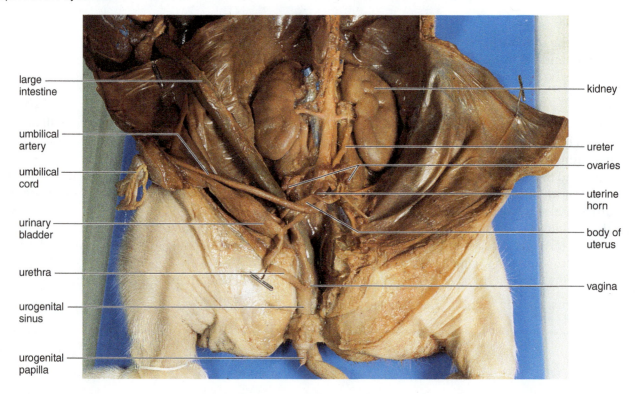

large intestine

umbilical artery

umbilical cord

urinary bladder

urethra

urogenital sinus

urogenital papilla

kidney

ureter

ovaries

uterine horn

body of uterus

vagina

Comparison of Female Fetal Pig with Human Female

Use Figure 20.12 to help you compare the female pig reproductive system with the human female reproductive system. Especially compare the anatomy of the oviducts in humans with that of the uterine horns in a pig. In a pig, the fetuses develop in the uterine horns; in a human female, the fetus develops in the body of the uterus. Complete Table 20.5, which compares the appearance of the oviducts and the uterus, as well as the presence or absence of a urogenital sinus in these two mammals.

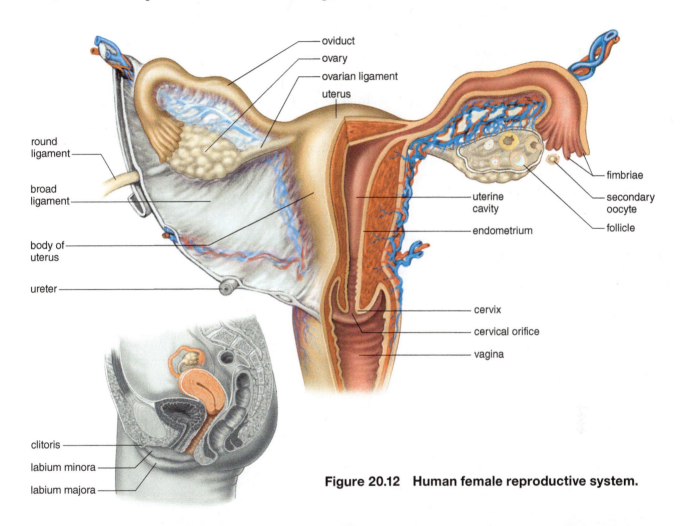

Figure 20.12   Human female reproductive system.

Table 20.5	Comparison of Female Fetal Pig with Human Female	
	Fetal Pig	**Human**
Oviducts		
Uterus		
Urogenital sinus		

Storage of Pigs

Before leaving the laboratory, dispose of or store your pig according to the directions of your instructor. Clean the dissecting tray and tools, and return them to their proper location. Wipe off your goggles and wash your hands.

1. What are the four chambers of the mammalian heart? _____

   _____

2. Contrast the pumping function of the right and left sides of the heart. _____

   _____

3. Sequence the blood vessels in the systemic circuit to trace the path of blood from the left ventricle to the

   kidneys and back to the right atrium. _____

   _____

4. Sequence the organs of the respiratory system from the glottis to the lungs. _____

   _____

5. What's the difference between the ureters and the urethra in the urinary system? ____

   _____

6. Sequence the following organs: stomach, large intestine, small intestine, pharynx, mouth, esophagus, and

   anus. _____

   _____

7. Sequence the path of sperm from the testes to the urogenital opening. _____

   _____

8. What organs enter the urogenital sinus in female pigs? _____

9. Which organ in males produces sperm, and which organ in females produces eggs? _____

   _____

10. How and when do sperm acquire access to an egg in mammals? _____

   _____

Essentials of Biology Website

Instructors can find lab prep information and answers to all of the laboratory questions in the Laboratory Resource Guide. *Students* can practice their knowledge with quizzes, animations, flashcards, and much more.

www.mhhe.com/maderessentials4

McGraw-Hill Access Science Website

An online encyclopedia of science and technology that provides information, including videos, that can enhance the laboratory experience.

www.accessscience.com

21

The Nervous System and Senses

Introduction

The vertebrate nervous system has two major divisions: the central nervous system (CNS), consisting of the brain and spinal cord, and the peripheral nervous system (PNS), which contains cranial nerves and spinal nerves (Fig. 21.1). Sensory receptors detect changes in environmental stimuli, and nerve impulses move along sensory nerve fibers to the brain and the spinal cord. The brain and spinal cord sum up the data before sending impulses via motor nerve fibers to effectors (muscles and glands) so a response to stimuli is possible. Nervous tissue consists of neurons; whereas the brain and spinal cord contain all parts of neurons, nerves contain only axons.

Figure 21.1   The nervous system.
The central nervous system (CNS) is in the midline of the body, and the peripheral nervous system (PNS) is outside the CNS.

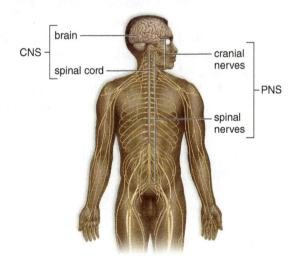

21.1 Central Nervous System

The brain is the enlarged anterior end of the spinal cord; it contains centers that receive input from and can command other regions of the nervous system.

> ⚠️ **Latex gloves** Wear protective latex gloves when handling preserved animal organs. Use protective eyewear and exercise caution when using sharp instruments during this laboratory. Wash hands thoroughly upon completion of this laboratory.

Preserved Sheep Brain

The sheep brain (Fig. 21.2) is often used to study the brain. It is easily available and large enough that individual parts can be identified.

Observation: Preserved Sheep Brain

Examine the exterior and a midsagittal (longitudinal) section of a preserved sheep brain or a model of the human brain, and with the help of Figure 21.1, identify the following:

1. **Ventricles:** Interconnecting spaces that produce and serve as a reservoir for cerebrospinal fluid, which cushions the brain. Toward the anterior, note the lateral ventricle (on one longitudinal section) and similarly a lateral ventricle (on the other longitudinal section). Trace the second ventricle to the third and then the fourth ventricles.
2. **Cerebrum:** Most developed area of the brain; responsible for higher mental capabilities. The cerebrum is divided into the right and left **cerebral hemispheres,** joined by the **corpus callosum,** a broad sheet of white matter. The outer portion of the cerebrum is highly convoluted and divided into the following surface lobes (see Fig. 21.2):
   a. **Frontal lobe:** Controls motor functions and permits voluntary muscle control; it also is responsible for abilities to think, problem solve, speak, and smell.
   b. **Parietal lobe:** Receives information from sensory receptors located in the skin and the taste receptors in the mouth. A groove called the **central sulcus** separates the frontal lobe from the parietal lobe.
   c. **Occipital lobe:** Interprets visual input and combines visual images with other sensory experiences. The optic nerves split and enter opposite sides of the brain at the optic chiasma, located in the diencephalon.
   d. **Temporal lobe:** Has sensory areas for hearing and smelling. The olfactory bulb contains nerve fibers that communicate with the olfactory cells in the nasal passages and take nerve impulses to the temporal lobe.
3. **Diencephalon:** Portion of the brain where the third ventricle is located. The hypothalamus and thalamus are also located here.
   a. **Thalamus:** Two connected lobes located in the roof of the third ventricle. The thalamus is the highest portion of the brain to receive sensory impulses before the cerebrum. It is believed to control which received impulses are passed on to the cerebrum. For this reason, the thalamus sometimes is called the "gatekeeper to the cerebrum."
   b. **Hypothalamus:** Forms the floor of the third ventricle and contains control centers for appetite, body temperature, blood pressure, and water balance. Its primary function is homeostasis. The hypothalamus also has centers for pleasure, reproductive behavior, hostility, and pain.
4. **Cerebellum:** Located just posterior to the cerebrum as you observe the brain dorsally, the cerebellum's two lobes make it appear rather like a butterfly. In cross section, the cerebellum has an internal pattern that looks like a tree. The cerebellum coordinates equilibrium and motor activity to produce smooth movements.

Figure 21.2 The sheep brain.

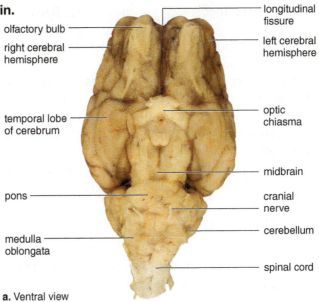

olfactory bulb

right cerebral hemisphere

temporal lobe of cerebrum

pons

medulla oblongata

longitudinal fissure

left cerebral hemisphere

optic chiasma

midbrain

cranial nerve

cerebellum

spinal cord

a. Ventral view

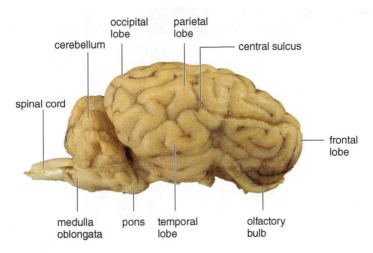

occipital lobe

parietal lobe

central sulcus

cerebellum

spinal cord

frontal lobe

medulla oblongata

pons

temporal lobe

olfactory bulb

b. Lateral view

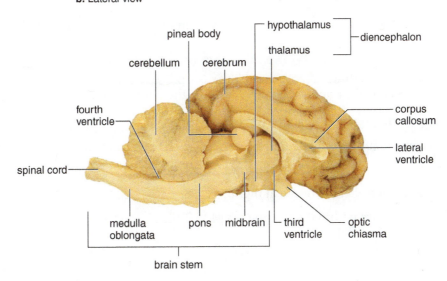

hypothalamus

diencephalon

pineal body

thalamus

cerebellum

cerebrum

fourth ventricle

corpus callosum

lateral ventricle

spinal cord

medulla oblongata

pons

midbrain

third ventricle

optic chiasma

brain stem

c. Longitudinal section

5. **Brain stem:** Part of the brain that connects with the spinal cord. Because it includes the pons and medulla oblongata, it contains centers for the functioning of internal organs; because of its location, it serves as a relay station for nerve impulses passing from the cord to the brain. Therefore, it helps keep the rest of the brain alert and functioning.

   a. **Midbrain:** Anterior to the pons, the midbrain serves as a relay station for sensory input and motor output. It also contains a reflex center for eye muscles.

   b. **Pons:** The ventral, bulblike enlargement on the brain stem. It serves as a passageway for nerve impulses running between the medulla and the higher brain regions.

   c. **Medulla oblongata** (or simply **medulla**): The most posterior portion of the brain stem. It controls internal organs; for example, blood pressure, cardiac, and breathing control centers are present in the medulla. Nerve impulses pass from the spinal cord through the medulla to and from higher brain regions.

The Human Brain

Based on your knowledge of the sheep brain, complete Table 21.1 by stating the major functions of each part of the brain listed. *Also label the human brain in Figure 21.3.*

Table 21.1	Summary of Brain Functions
Part	**Major Functions**
Cerebrum	
Cerebellum	
Diencephalon Thalamus	
Hypothalamus	
Brain stem Midbrain	
Pons	
Medulla oblongata	

Which parts of the brain work together to achieve the following?

1. Good eye–hand coordination _____

2. Concentrating on homework when TV is playing _____

3. Avoiding dark alleys while walking home at night _____

4. Keeping the blood pressure within the normal range _____

Comparison of Vertebrate Brains

The vertebrate brain has a forebrain, midbrain, and hindbrain. In the earliest vertebrates, the forebrain was largely a center for sense of smell, the midbrain was a center for the sense of vision, and the hindbrain was a center for the sense of hearing and balance. How the functions of these parts changed to accommodate the lifestyles of different vertebrates can be traced.

Observation: Comparison of Vertebrate Brains

1. Examine the brain of a reptile, bird, and mammal (Fig. 21.3).
2. Identify the following three areas, and for each animal featured.

   a. **Forebrain:** Contains olfactory bulb and cerebrum
   b. **Midbrain:** Contains optic lobe (and other structures)
   c. **Hindbrain:** Contains cerebellum (and other structures)

In fishes and amphibians, the midbrain is often the most prominent region of the brain because it contains higher control centers. In these animals, the cerebrum largely has an olfactory function. Fishes, birds, and mammals have a well-developed cerebellum, which may be associated with these animals' agility. In reptiles, birds, and mammals, the cerebrum becomes increasingly complex. This may be associated with the cerebrum's increasing control over the rest of the brain and the evolution of areas responsible for thought and reasoning.

Figure 21.3 Vertebrate brains.
a. A comparison of reptile, bird, and mammalian brains shows that the forebrain increased in size and complexity among these animals.
b. The human brain.

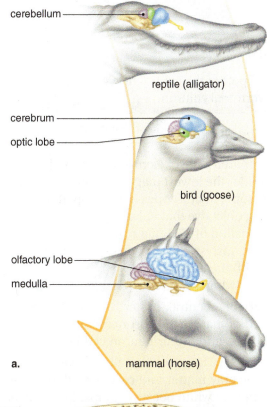

reptile (alligator)

cerebellum

cerebrum
optic lobe

bird (goose)

olfactory lobe
medulla

a.

mammal (horse)

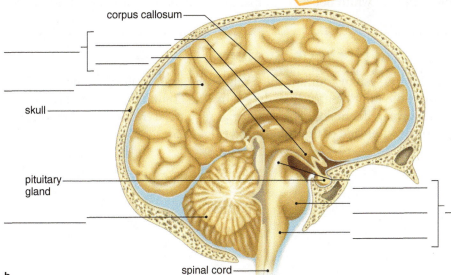

corpus callosum

skull

pituitary gland

spinal cord

b.

Observe any available vertebrate brains and record your observations in Table 21.2.

Table 21.2 Comparison of Vertebrate Brains

Vertebrate	Observations
Fish (shark)	
Reptile (alligator)	
Bird (goose)	
Mammal (horse)	

The Spinal Cord

The spinal cord is a part of the central nervous system. It lies in the middorsal region of the body and is protected by the vertebral column.

Observation: The Spinal Cord

1. Examine a prepared slide of a cross section of the spinal cord under the lowest magnification possible. For example, some microscopes are equipped with a short scanning objective that enlarges about 3.5×, with a total magnification of 35×. If a scanning objective is not available, observe the slide against a white background with the naked eye.

2. Identify the following with the help of Figure 21.4:

    a. **Gray matter:** A central, butterfly-shaped area composed of masses of short nerve fibers, interneurons, and motor neuron cell bodies

    b. **White matter:** Masses of long fibers that lie outside the gray matter and carry impulses up and down the spinal cord. In animals, white matter appears white because an insulating myelin sheath surrounds long fibers.

Figure 21.4   The spinal cord.
Photomicrograph of spinal cord cross section.

central canal

gray matter

white matter

20×

21.2   Peripheral Nervous System

The peripheral nervous system contains the cranial nerves and the spinal nerves. Twelve pairs of cranial nerves project from the inferior surface of the brain. The cranial nerves are largely concerned with nervous communication between the head, neck, and facial regions of the body and the brain. The 31 pairs of spinal nerves emerge from either side of the spinal cord (Fig. 21.5).

Spinal Nerves

Each spinal nerve contains long fibers of sensory neurons and long fibers of motor neurons. In Figure 21.5, identify the following:

1. **Sensory neuron:** Takes nerve impulses from a sensory receptor to the spinal cord. The cell body of a sensory neuron is in the dorsal root ganglion.
2. **Interneuron:** Lies completely within the spinal cord. Some interneurons have long fibers and take nerve impulses to and from the brain. The interneuron in Figure 21.5 transmits nerve impulses from the sensory neuron to the motor neuron.
3. **Motor neuron:** Takes nerve impulses from the spinal cord to an effector—in this case, a muscle. Muscle contraction is one type of response to stimuli.

Suppose you were walking barefoot and stepped on a prickly sandbur. Describe the pathway of information, starting with the pain receptor in your foot, that would allow you to both feel and respond to this unwelcome stimulus. _____

_____

_____

Spinal Reflexes

A **reflex** is an involuntary and predictable response to a given stimulus that allows a quick response to environmental stimuli without communicating with the brain. When you touch a sharp pin, you immediately withdraw your hand (see Fig. 21.5). When a spinal reflex occurs, a sensory receptor is

Figure 21.5 Spinal nerves and spinal cord.
The arrows mark the path of nerve impulses from a sensory receptor to an effector.

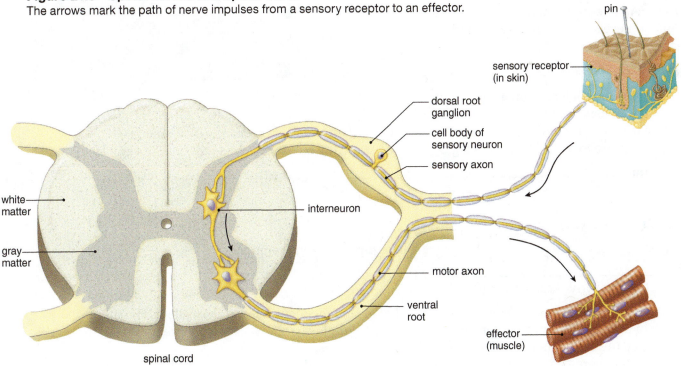

stimulated and generates nerve impulses, which pass along the three neurons mentioned earlier—the sensory neuron, interneuron, and motor neuron—until the effector responds. In the spinal reflexes that follow, a receptor detects the pin, and sensory neurons conduct nerve impulses to interneurons in the spinal cord. The interneurons send a message via motor neurons to the effectors, muscles in the leg or foot. These reflexes are involuntary because the brain is not involved in formulating the response. *Consciousness* of the stimulus lags behind the response because information must be sent up the spinal cord to the brain before you can become aware of the pin.

Experimental Procedure: Spinal Reflex

Although many reflexes occur in the body, only a tendon reflex is investigated in this Experimental Procedure. One easily tested tendon reflex involves the **patellar tendon.** When this tendon is tapped with a reflex hammer (Fig. 21.6) or, in this experiment, with a meterstick, the attached muscle is stretched. This causes a receptor to generate nerve impulses, which are transmitted along sensory neurons to the spinal cord. Nerve impulses from the cord then pass along motor neurons and stimulate the muscle, causing it to contract. As the muscle contracts, it tugs on the tendon, causing movement of a bone opposite the joint. Receptors in other tendons, such as the Achilles tendon, respond similarly. Such reflexes help the body automatically maintain balance and posture.

Figure 21.6 Knee-jerk reflex.
The quick response when the patellar tendon is stimulated by tapping with a reflex hammer indicates that a reflex has occurred.

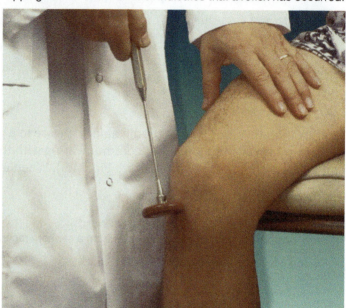

Knee-jerk (patellar) reflex

Knee-Jerk (Patellar) Reflex

1. Have the subject sit on a table so that his or her legs hang freely.
2. Sharply tap one of the patellar tendons just below the patella (kneecap) with a meterstick.
3. In this relaxed state, does the leg flex (move toward the buttocks) or extend (move away from the buttocks)? _____

21.3 Animal Eyes

The eye is a special sense organ for detecting light rays in the environment.

Anatomy of Invertebrate Eyes

Arthropods have **compound eyes** composed of many independent visual units, called ommatidia, each of which has its own photoreceptor cells (Fig. 21.7*a*). Each unit "sees" a separate portion of the object. How well the brain combines this information is not known.

A chambered nautilus has a camera type of eye. A single lens focuses an image of the visual field on the photoreceptors, packed closely together (Fig. 21.7*b*).

Figure 21.7 The compound eye compared to the camera-type eye.

a. A fly has a compound eye. Each visual unit of a compound eye has a cornea and a lens that focuses light onto photoreceptor cells. These cells generate nerve impulses transmitted to the brain, where interpretation produces a mosaic image. **b.** A cephalopod, such as a chambered nautilus and a squid, have a camera-type eye similar to those of vertebrates.

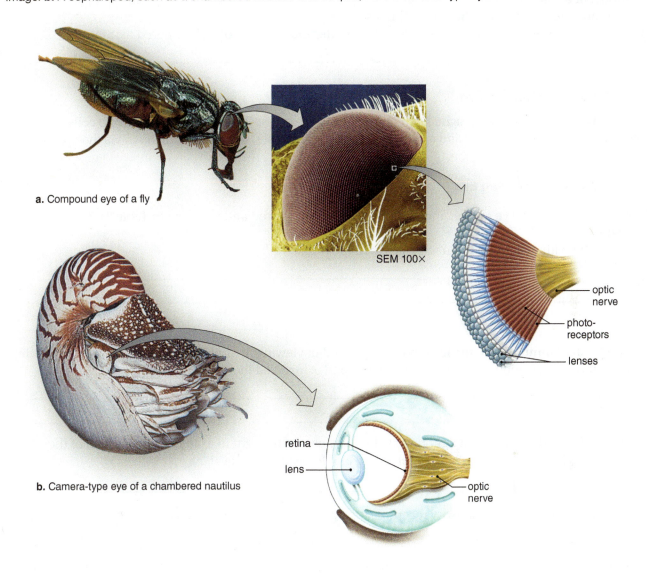

a. Compound eye of a fly

SEM 100×

optic nerve

photo-receptors

lenses

b. Camera-type eye of a chambered nautilus

retina

lens

optic nerve

1. Examine the demonstration slide of a compound eye set up under a stereomicroscope, or examine a model of a compound eye.
2. Examine the eyes of any other invertebrates on display.
3. Explain why the eye of a fly can detect motion better than the eye of a chambered nautilus. _____

_____

Anatomy of the Human Eye

The human eye is responsible for sight. Light rays enter the eye and strike the **rod cells** and **cone cells,** the photoreceptors for sight. The rods and cones generate nerve impulses, which go to the brain via the optic nerve.

Observation: The Human Eye

1. Examine a human eye model, and identify the structures listed and depicted in Table 21.3 and Figure 21.8.
2. Trace the path of light from outside the eye to the retina.

   _____

   _____

   _____

3. During **accommodation,** the lens rounds up to aid in viewing near objects or flattens to aid in viewing distant objects. Which structure holds the lens and is involved in accommodation?

   _____

   _____

4. **Refraction** is the bending of light rays so that they can be brought to a single focus. Which of the structures listed in Table 21.3 aid in refracting and focusing light rays?

   _____

5. Specifically, what are the sensory receptors for sight, and where are they located in the eye?

   _____

   _____

6. What structure takes nerve impulses to the brain from the rod cells and cone cells?

   _____

7. Which cerebral lobe processes nerve impulses from an eye? _____

Table 21.3 Parts of the Human Eye

Part	Location	Function
Sclera	Outer layer of eye	Protects and supports eyeball
Cornea	Transparent portion of sclera	Refracts light rays
Choroid	Middle layer of eye	Absorbs stray light rays
Retina	Inner layer of eye	Contains receptors for sight
Rod cells	In retina	Make black-and-white vision possible
Cone cells	Concentrated in fovea centralis	Make color vision possible
Fovea centralis	Special region of retina	Makes acute vision possible
Lens	Interior of eye between cavities	Refracts and focuses light rays
Ciliary body	Extension from choroid	Holds lens in place; functions in accommodation
Iris	More anterior extension of choroid	Regulates light entrance
Pupil	Opening in middle of iris	Admits light
Humors (aqueous and vitreous)	Fluid media in anterior and posterior compartments, respectively, of eye	Transmit and refract light rays; support eyeball
Optic nerve	Extension from posterior of eye	Transmits impulses to occipital lobe of brain

Figure 21.8 Anatomy of the human eye.
The sensory receptors for vision are the rod cells and cone cells in the retina of the eye.

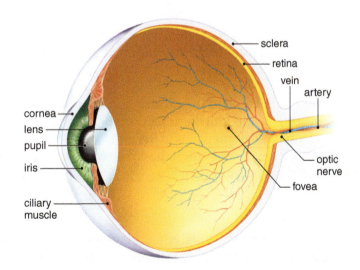

The Blind Spot of the Eye

The **blind spot** occurs where the optic nerve fibers exit the retina. No vision is possible at this location because of the absence of rod cells and cone cells.

This Experimental Procedure requires a laboratory partner. Figure 21.9 shows a small circle and a cross several centimeters apart.

Figure 21.9   Blind spot.
This dark circle (or cross) will disappear at one location because there are no rod cells or cone cells at each eye's blind spot, where vision does not occur.

Left Eye

1. Hold Figure 21.9 approximately 30 cm from your eyes. The cross should be directly in front of your left eye. If you wear glasses, keep them on.
2. Close your right eye.
3. Stare only at the cross with your left eye. You should also be able to see the circle in the same field of vision. Slowly move the paper toward you until the circle disappears.
4. Repeat the procedure as many times as needed to find the blind spot.
5. Then slowly move the paper closer to your eyes until the circle reappears. Because only your left eye is open, you have found the blind spot of your left eye.
6. With your partner's help, measure the distance from your eye to the paper when the circle first

   disappeared. Left eye: _____ cm

Right Eye

1. Hold Figure 21.9 approximately 30 cm from your eyes. The circle should be directly in front of your right eye. If you wear glasses, keep them on.
2. Close your left eye.
3. Stare only at the circle with your right eye. You should also be able to see the cross in the same field of vision. Slowly move the paper toward you until the cross disappears.
4. Repeat the procedure as many times as needed to find the blind spot.
5. Then slowly move the paper closer to your eyes until the cross reappears. Because only your right eye is open, you have found the blind spot of your right eye.
6. With your partner's help, measure the distance from your eye to the paper when the cross first

   disappeared. Right eye: _____ cm

Why are you unaware of a blind spot under normal conditions? Although the eye detects patterns of light and color, it is the brain that determines what we visually perceive. The brain interprets the visual input based in part on past experiences. In this exercise, you created an artificial situation in which you became aware of how your perception of the world is constrained by the eye's anatomy.

Accommodation of the Eye

When the eye accommodates to see objects at different distances, the shape of the lens changes. The lens shape is controlled by the ciliary muscles attached to it. When you are looking at a distant object, the lens is in a flattened state. When you are looking at a closer object, the lens becomes more rounded. The elasticity of the lens determines how well the eye can accommodate. Lens elasticity decreases with increasing age, a condition called **presbyopia.** Presbyopia is the reason many older people need bifocals to see near objects.

Experimental Procedure: Accommodation of the Eye

This Experimental Procedure requires a laboratory partner. It tests accommodation of either your left or your right eye.

1. Hold a pencil upright by the eraser and at arm's length in front of whichever of your eyes you are testing (Fig. 21.10).
2. Close the opposite eye.
3. Move the pencil from arm's length toward your eye.
4. Focus on the end of the pencil.
5. Move the pencil toward you until the end is out of focus. Measure the distance (in centimeters)

    between the pencil and your eye: _____ cm

6. At what distance can your eye no longer

    accommodate for distance? _____ cm

7. If you wear glasses, repeat this experiment without your glasses, and note the accommodation

    distance of your eye without glasses:

    _____ cm. (Contact lens wearers need not make these determinations, and they should write the words "contact lens" in this blank.)

8. The "younger" lens can easily accommodate for closer distances. The nearest point at which the end of the pencil can be clearly seen is called the **near point.** The more elastic the lens, the "younger" the eye

    (Table 21.4). How "old" is the eye you tested? _____

Figure 21.10 Accommodation.
When testing the ability of your eyes to accommodate to see a near object, always keep the pencil in this position.

Table 21.4 Near Point and Age Correlation						
Age (years)	10	20	30	40	50	60
Near point (cm)	9	10	13	18	50	83

21.4 Animal Ears

Among invertebrates, only certain arthropod groups—crustaceans, spiders, and insects—have receptors for detecting sound waves. If available, locate the tympanum of a grasshopper (Fig. 21.11a). The tympanum covers an internal air sac that allows the tympanum to vibrate when stuck by sound waves. Sensory neurons attached to the tympanum are stimulated directly by the vibration.

Anatomy of Human Ear

The human ear may have evolved from the lateral line of fishes. It has a series of hair cells with cilia embedded in a mass of gelatinous material that detects water currents and pressure waves from nearby objects (Fig. 21.11b). Hair cells in the human ear have hair cells with stereocilia embedded in the gelatinous tectorial membrane. When the basal membrane vibrates, bending of the stereocilia generates nerve impulses that result in hearing (Fig. 21.11c).

Figure 21.11 Evolution of the human ear.
a. A few invertebrates have a tympanum to receive sound waves as in a grasshopper. **b.** The lateral line of fishes which detects vibrations and movement contains hair cells with cilia embedded in a gelatinous cupula. **c.** Hair cells in the cochlea of the human ear have stereocilia embedded in the gelatinous tectorial membrane. When the basal membrane vibrates, bending of the stereocilia generates nerve impulses that result in hearing.

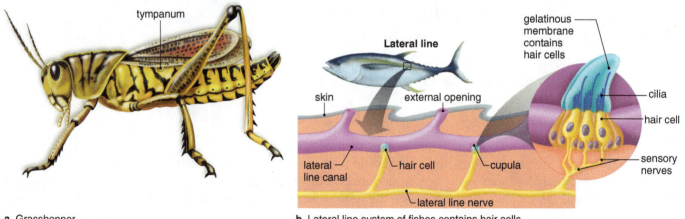

a. Grasshopper

b. Lateral line system of fishes contains hair cells.

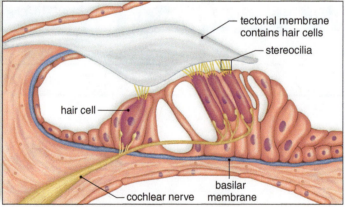

c. The cochlea of the human ear contains hair cells.

The human ear, whose parts are listed and depicted in Table 21.5 and Figure 21.12, serves two functions: hearing and balance.

Observation: The Human Ear

Examine a human ear model, and find the structures depicted in Figure 21.12 and listed in Table 21.5.

Table 21.5 Parts of the Human Ear			
Part	**Medium**	**Function**	**Mechanoreceptor**
Outer ear	Air		
Pinna		Collects sound waves	—
Auditory canal		Filters air	—
Middle ear	Air		
Tympanic membrane and ossicles		Amplify sound waves	—
Auditory tube		Equalizes air pressure	—
Inner ear	Fluid		
Semicircular canals		Rotational equilibrium	Stereocilia embedded in cupula
Vestibule (contains utricle and saccule)		Gravitational equilibrium	Stereocilia embedded in otolithic membrane
Cochlea (spiral organ)		Hearing	Stereocilia embedded in tectorial membrane

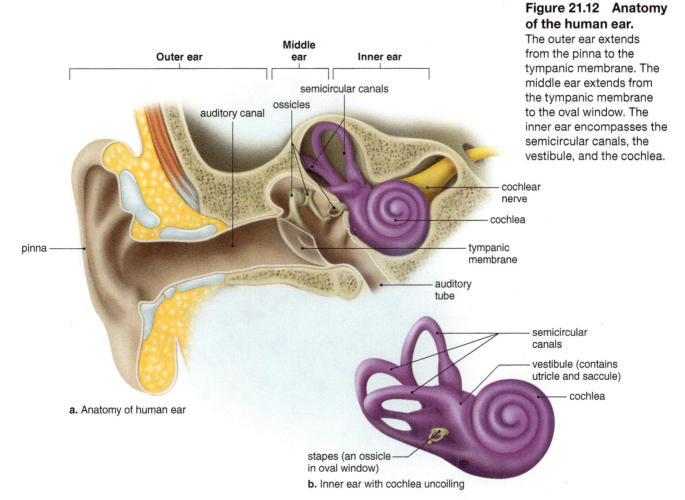

Figure 21.12 Anatomy of the human ear.
The outer ear extends from the pinna to the tympanic membrane. The middle ear extends from the tympanic membrane to the oval window. The inner ear encompasses the semicircular canals, the vestibule, and the cochlea.

a. Anatomy of human ear

b. Inner ear with cochlea uncoiling

Physiology of the Human Ear

When you hear, sound waves are picked up by the **tympanic membrane** and amplified by the **malleus, incus,** and **stapes.** This creates pressure waves in the canals of the **cochlea,** which lead to stimulation of **hair cells,** the receptors for hearing. Hair cells in the utricle and saccule of the vestibule and in semicircular canals are receptors for equilibrium (i.e., balance). Nerve impulses from the cochlea travel by way of the cochlear nerve and the vestibular nerve to the brain and eventually are interpreted by the brain as sound.

Experimental Procedure: Locating Sound

Humans locate the direction of sound according to how well it is detected by either or both ears. A difference in the hearing ability of the two ears can lead to a mistaken judgment about the direction of sound. You and a laboratory partner should perform this Experimental Procedure on each other.

1. Ask the subject to be seated, with eyes closed. Then strike a tuning fork or rap two spoons together at the five locations listed in number 2. Use a random order.

2. Ask the subject to give the exact location of the sound in relation to his or her head. Record the subject's perceptions when the sound is

   a. directly below and behind the head. _____

   b. directly behind the head. _____

   c. directly above the head. _____

   d. directly in front of the face. _____

   e. to the side of the head. _____

3. Is there an apparent difference in hearing between the subject's two ears? _____

21.5   Sensory Receptors in Human Skin

The sensory receptors in human skin respond to touch, pain, temperature, and pressure (Fig. 21.13). There are individual sensory receptors for each of these stimuli, as well as free nerve endings able to respond to pressure, pain, and temperature.

Figure 21.13   Sensory receptors in the skin.
Each type of receptor shown responds primarily to a particular stimulus.

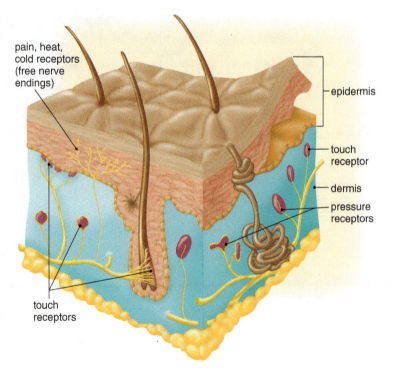

pain, heat, cold receptors (free nerve endings)

epidermis

touch receptor

dermis

pressure receptors

touch receptors

The dermis of the skin contains touch and temperature receptors.

Sense of Touch

You will need a laboratory partner to perform this Experimental Procedure. Enter *your* data, not the data of your partner, in the spaces provided.

1. Ask the subject to be seated, with eyes closed. You are going to test the subject's ability to discriminate between the two points of a hairpin or a pair of scissors at the four locations noted below.
2. Hold the points of the hairpin or scissors on the given skin area, with both points gently touching the subject. Ask the subject whether the experience involves one or two touch sensations.
3. Record the shortest distance between the hairpin or scissor points for a two-point discrimination.

   a. Forearm: _____ mm

   b. Back of the neck: _____ mm

   c. Index finger: _____ mm

   d. Back of the hand: _____ mm

4. Which of these areas apparently contains the greatest density of touch receptors? _____

   Why is this useful? _____

5. Do you have a sense of touch at every point in your skin? _____ Explain. _____

6. What specific part of the brain processes nerve impulses from touch and pain receptors?

Sense of Heat and Cold

1. Obtain three 1,000 ml beakers, and fill one with *ice water,* one with *tap water* at room temperature, and one with *warm water* (45°–50°C).
2. Immerse your left hand in the ice-water beaker and your right hand in the warm-water beaker for 30 seconds.
3. Then place both hands in the beaker with room-temperature tap water.
4. Record the sensation in the right and left hands.

   a. Right hand: _____

   b. Left hand: _____

5. Explain your results. _____

21.6 Human Chemoreceptors

The taste receptors, called _____, located in the mouth, and the smell receptors, called _____, located in the nasal cavities, are the chemoreceptors that respond to molecules in the air and water. Nerve impulses from taste receptors go to the _____ lobe of the brain, while those from smell receptors go to the _____ lobe of the brain.

Experimental Procedure: Sense of Taste and Smell

You will need a laboratory partner to perform the following procedures. It will not be necessary for all tests to be performed on both partners. You should take turns being either the subject or the experimenter. Dispose of used cotton swabs in a hazardous waste container or as directed by your instructor.

Taste and Smell

1. Students work in groups. Each group has one experimenter and several subjects.
2. The experimenter should obtain a LifeSavers candy from the various flavors available, without letting the subject know what flavor it is.
3. The subject closes both eyes and holds his or her nose.
4. The experimenter gives the LifeSavers candy to the subject, who places it on his or her tongue.
5. The subject, while still holding his or her nose, guesses the flavor of the candy. The experimenter records the guess in Table 21.6.
6. The subject releases his or her nose and guesses the flavor again. The experimenter records the guess and the actual flavor in Table 21.6.

Table 21.6 Taste and Smell Experiment			
Subject	Actual Flavor	Flavor While Holding Nose	Flavor After Releasing Nose
1			
2			
3			

Conclusions: Sense of Taste and Smell

- From your results, how would you say that smell affects the taste of LifeSavers candy?

  _____

- What do you conclude about the effect of smell on your sense of taste?

  _____

1. Describe the cerebrum of the human brain, and state a function. _____
   _____

2. The brain stem includes the medulla oblongata, the pons, and the midbrain. Explain the expression *brain stem* as an anatomical term. _____
   _____

3. Describe the location of the gray/white matter of the spinal cord, and give a function for each.
   _____
   _____

4. State, in order from receptor to effector, the neurons associated with a spinal reflex. _____
   _____

5. Trace the path of light in the human eye—from the exterior to the retina and then from retinal nerve impulses to the brain. _____
   _____

6. Contrast the eye of an arthropod with the eye of a chambered nautilus and human. _____
   _____

7. If you move an illustration that contains a dark circle and a dark cross toward an eye, one or the other may disappear. Give an explanation for this. _____
   _____

8. Trace the path of sound waves in the human ear—from the tympanic membrane to the receptors for hearing. _____
   _____

9. Compare the manner in which a grasshopper "hears" to the way a human hears.
   _____
   _____

10. Name four structures located in the dermis of the skin. _____
    _____

Essentials of Biology Website

Instructors can find lab prep information and answers to all laboratory questions in the Laboratory Resource Guide. *Students* can practice their knowledge with quizzes, animations, flashcards, and much more.

www.mhhe.com/maderessentials4

McGraw-Hill Access Science Website

An online encyclopedia of science and technology that provides information, including videos, that can enhance the laboratory experience.

www.accessscience.com

22
Effects of Pollution on Ecosystems

Learning Outcomes

22.1 Studying the Effects of Pollutants
- Predict the effect of oxygen deprivation on species composition and diversity of ecosystems.
- Predict the effect of acid deposition on species composition and diversity of ecosystems.
- Predict the effect of enrichment on species composition and diversity of ecosystems.

22.2 Studying the Effects of Cultural Eutrophication
- Identify potential sources of pollution, and predict the effect of cultural eutrophication on the food chains of a lake or pond in your vicinity.

Introduction

Pollution is an environmental change brought about by humans that adversely affects the health of humans and all other organisms. This laboratory will consider three causes of aquatic pollution: thermal pollution, acid pollution, and cultural eutrophication. **Thermal pollution** occurs when water temperature rises above normal. Deforestation, soil erosion, and the burning of fossil fuels contribute to thermal pollution. The chief cause is use of water from a lake or the ocean as a coolant for the excess heat of a power plant. As water temperature rises, the amount of dissolved oxygen decreases, and most cells require a goodly supply of oxygen. **Acid deposition** (acid rain or snow) occurs after sulfur dioxide and nitrogen oxide enter the atmosphere, usually from the burning of fossil fuels. These gases are converted to acids, which return to Earth. Acid deposition kills small aquatic invertebrates and decomposers, so that the entire ecosystem is threatened.

   Cultural eutrophication is another type of pollution caused by humans. It occurs when excess chemicals enter bodies of water from agricultural fields, wastewater from sewage treatment plants, or even excess detergents. These sources of excess nutrients cause an algal bloom, seen as a green scum on a lake. When aquatic populations overgrow and die, decomposition robs the lake of oxygen, the fish die off, and the water is unfit for human consumption.

22.1 Studying the Effects of Pollutants

We are going to study the effects of pollution by observing its effects on hay infusion microorganisms (Fig. 22.1), on seed germination, and on an animal called *Gammarus*.

Study of Hay Infusion Cultures

A hay infusion culture (hay soaked in water) contains various microscopic organisms, but we will be concentrating on how the pollutants in our study affect the protozoan populations in the culture. We will consider both of these aspects: **species composition,** the number of different types of protozoans, and **species diversity,** the relative abundance of each type of protozoan.

Experimental Procedure: Effect of Pollutants on a Hay Infusion Culture

During this Experimental Procedure you will microscopically examine samples of these hay infusion cultures: (1) no treatment, the control culture; (2) oxygen deprived, heating causes water to lose oxygen; (3) acidic culture, acid has been added; and (4) overenriched with nutrients, inorganic nutrients have been added. In each instance you will do the following.

Prepare a wet mount of the culture and report the results of your examination in Table 22.1 in terms of species composition and species diversity.

1. **Control culture:** This culture simulates the species composition and diversity of a nonpolluted environment. Prepare a wet mount and, with the assistance of Figure 22.1 and any guides available in the laboratory, identify as many different types of protozoans as possible in the hay infusion culture. State whether species composition is high, medium, or low. Record your estimation in the second column of Table 22.1. Do you judge species diversity to be high, medium, or low? Record your estimation in the third column of Table 22.1.

Figure 22.1 Protozoans in hay infusion cultures.
Organisms are not to size.

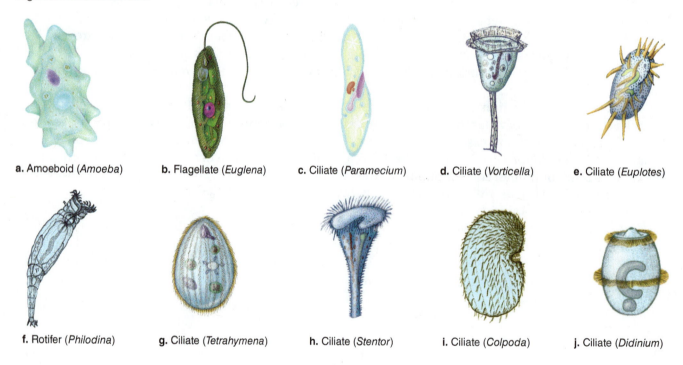

a. Amoeboid (*Amoeba*) b. Flagellate (*Euglena*) c. Ciliate (*Paramecium*) d. Ciliate (*Vorticella*) e. Ciliate (*Euplotes*)

f. Rotifer (*Philodina*) g. Ciliate (*Tetrahymena*) h. Ciliate (*Stentor*) i. Ciliate (*Colpoda*) j. Ciliate (*Didinium*)

Table 22.1 Effect of Pollution on a Hay Infusion Culture

Type of Culture	Species Composition (high, medium, or low)	Species Diversity (high, medium, or low)	Explanation
Control			
Oxygen-deprived			
Acidic			
Enriched			

2. **Oxygen-deprived culture** due to thermal pollution: In order for humans to generate electricity, power plants remove cool water from, and return warm water to, a nearby lake or ocean. When we study the effects of low oxygen on a hay infusion culture, we are studying an effect of thermal pollution, because warm water holds less oxygen than cold water.

   Prepare a wet mount of this culture and determine if there is a change in species composition and diversity. Again record the species composition and species diversity as high, medium, or low in Table 22.1.

3. **Acidic culture:** When humans burn fuels to heat their houses, to run any sort of vehicle, or to generate electricity, acids are a by-product that enter the atmosphere, where they cause acid rain. In this culture the pH has been adjusted to 4 with sulfuric acid (H_2SO_4). This simulates the effect of acid rain on a hay infusion culture.

   Prepare a wet mount of this culture and determine if there is a change in species composition and diversity. Again record the species composition and species diversity as high, medium, or low in Table 22.1.

4. **Enriched culture:** Organic nutrients enter water from sewage treatment plants—even detergents contain inorganic nutrients—and from agricultural fields and our lawns due to the use of fertilizers. These nutrients will cause the algae population, which is food for most protozoans, to increase. In the short term, their species composition should increase. Eventually, as the algae die off, decomposition will rob the water of oxygen and the protozoans may start to die off.

   Prepare a wet mount of this culture and determine if there is a change in species composition and diversity. Again record the species composition and species diversity as high, medium, or low in Table 22.1.

Conclusions: Effect of Pollution on a Hay Infusion Culture

- What could be a physiological reason for the adverse effects of oxygen deprivation on a hay infusion culture? If consistent with your results, enter this explanation in the last column of Table 22.1.
- What could be a physiological reason for the adverse effects of a low pH on a hay infusion culture? If consistent with your results, enter this explanation in the last column of Table 22.1.
- What could be a physiological reason for the adverse effects of an enriched culture? If consistent with your results, enter this explanation in the last column of Table 22.1.

Effect of Acid Rain on Seed Germination

Seeds depend on favorable environmental conditions of temperature, light, and moisture to germinate, grow, and reproduce. Like any other biological process, gernimation requires enzymatic reactions that can be adversely affected by an unfavorable pH. Acid rain adversely affects the health of plants, animals, and humans. For example, many forests are now dying, some lakes are devoid of fish, and humans possibly suffer more bronchial disorders (e.g., asthma) due to acid rain.

Experimental Procedure: Effect of Acid Rain on Seed Germination

In this Experimental Procedure, we will test whether there is a negative correlation between acidic water and germination. In other words, it is hypothesized that, as acidity increases, the more likely seeds will _____.

Your instructor has placed 20 sunflower seeds in each of five containers with water of increasing acidity: 0% vinegar (tap water), 1% vinegar, 5% vinegar, 20% vinegar, and 100% vinegar.

1. Test and record the pH of solutions having the vinegar concentrations noted above. Record the pH of each solution in Table 22.2.
2. Count the number of germinated sunflower seeds in each container, and complete Table 22.2.

Table 22.2   Effect of Increasing Acidity on Germination of Sunflower Seeds

Concentration of Vinegar	pH	Number of Seeds That Germinated	Percent Germination
0%			
1%			
5%			
20%			
100%			

Conclusions: Effect of Acidity on Germination of Sunflower Seeds

- As you know, each enzyme has an optimum pH. Explain why acid rain is expected to inhibit metabolism, and therefore seedling development. _____

  _____

- Do the data support or falsify your hypothesis? _____

Study of *Gammarus*

A small crustacean called *Gammarus* lives in ponds and streams (Fig. 22.2) where it feeds on debris, algae, and anything else smaller than itself, such as some of the protozoans in Figure 22.1. In turn, fish like to feed on *Gammarus.*

Experimental Procedure: Gammarus

- Add 25 ml of spring water to a beaker and record the pH of the water. _____ pH

- Add four *Gammarus* to the container. Do they all use their legs in swimming? _____

- Which legs are used in jumping and climbing? _____

- What do *Gammarus* do when they "bump" into each other? _____ _____

Figure 22.2   *Gammarus.*
Gammarus is a type of crustacean classified in a subphylum that also includes shrimp.

Control Sample

After observing *Gammarus,* decide what two behaviors are most often observed. During a 5-minute time span, total the amount of time spent doing each of these behaviors.

Behaviors	Amount of Time	Total Time
1. _____	_____	_____
2. _____	_____	_____
3. _____	_____	_____

Test Sample

If so directed by your instructor, put a *Gammarus* in a beaker of spring water adjusted to pH 4 by adding vinegar. During a 5-minute time span, total the amount of time spent doing each of the same behaviors.

Behaviors	Amount of Time	Total Time
1. _____	_____	_____
2. _____	_____	_____
3. _____	_____	_____

Conclusions

- Draw a conclusion from this study. _____ _____ _____

- Create a food chain that shows who eats whom and includes algae, protozoans, *Gammarus,* fish, and humans. _____

  What would happen to this food chain if the water were

  oxygen deprived? _____

  Acidic? _____

  Enriched with inorganic nutrients (short term and long term)? _____

  _____

  _____

Conclusions: Studying the Effects of Pollutants

- Give an example to show that the hay infusion study pertains to real ecosystems. _____

  _____

- What are the potential consequences of acid rain on crops that reproduce by seeds? _____

  _____

  On the food chains of the ocean? _____

- How does the addition of nutrients affect species composition and species diversity of an ecosystem over time? _____

22.2   Studying the Effects of Cultural Eutrophication

Chlorella, the green alga used in this study, is considered to be representative of algae in bodies of fresh water. The protozoan *Daphnia* (Fig. 22.3) feeds on green algae, such as *Chlorella.* First, you will observe how *Daphnia* feeds, and then you will determine the extent to which *Daphnia* could keep the effects of cultural eutrophication from occurring in a hypothetical example. Keep in mind that this case study is an oversimplification of a generally complex problem.

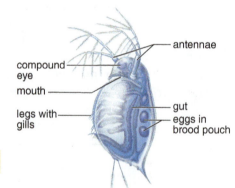

Figure 22.3   Anatomy of *Daphnia*, an arthropod.

Observation: Daphnia *Feeding*

1. Place a small pool of petroleum jelly in the center of a small petri dish.
2. Use a dropper to take a *Daphnia* from the stock culture, place it on its back (covered by water) in the petroleum jelly, and observe it under the stereomicroscope.
3. Note the clamlike carapace and the legs waving rapidly as the *Daphnia* filters the water.
4. Add a drop of carmine solution, and observe how the *Daphnia* filters the "food" from the water and passes it through the gut. The gut is more visible if you push the animal onto its side. In this position, you may also observe the heart beating in the region above the gut and just behind the head.
5. Allow the *Daphnia* to filter-feed for up to 30 minutes, and observe the progress of the carmine particles through the gut. Does the carmine travel completely through the gut in 30 minutes?

   _____

This Experimental Procedure requires the use of a spectrophotometer. Absorbance will be a measure of the algal population level; the greater the number of algal cells, the greater the absorbance. The higher the absorbance, the greater the amount of light absorbed and *not* passed through the solution.

1. Obtain two spectrophotometer tubes (cuvettes) and a Pasteur pipet.
2. Fill one of the cuvettes with distilled water, and use it to zero the spectrophotometer. Save this tube for number 6.
3. Use the Pasteur pipet to fill the second cuvette with *Chlorella*. Gently aspirate and expel the sample several times (without creating bubbles) to give a uniform dispersion of the algae.
4. Add ten hungry *Daphnia,* and following your instructor's directions, immediately measure the absorbance with the spectrophotometer. If a *Daphnia* swims through the beam of light, a strong deflection should occur; do not use any such higher readings—instead, use the lower reading for the absorbance. Record your reading here. _____
5. Remove the cuvette with the *Daphnia* to a safe place in a test tube rack. Allow the *Daphnia* to feed for 30 minutes.
6. Rezero the spectrophotometer with the distilled water cuvette.
7. Measure the absorbance of the experimental cuvette again. Record your reading here and explain why the absorbance changed. _____

_____ .

The following problem will test your understanding of the value of a single species—in this case, *Daphnia*. Please realize that this is an oversimplification of a generally complex problem.

1. Assume that developers want to build condominium units on the shores of Silver Lake. Home-owners in the area have asked the regional council to determine how many units can be built without causing the lake to become eutrophic. As a member of the council, you have been given the following information.

   The current population of *Daphnia,* 10 animals/liter, presently filters 24% of the lake per day, meaning that it removes this percentage of the algal population per day. This is sufficient to keep the lake essentially clear. Predation—the eating of the algae—will allow the *Daphnia* population to increase to no more than 50 animals/liter. Therefore, 50 *Daphnia*/liter will be available for feeding on the increased number of algae that would result from building the condominiums.

   Using this information, complete Table 22.3.

Table 22.3   *Daphnia* Filtering

Number of *Daphnia*/Liter	Percent of Lake Filtered
10	24%
50	

2. The sewage system of the condominium will add nutrients to the lake. Phosphorous output will be 1 kg per day for every ten condominiums. This will cause a 30% increase in the algal population. Using this information, complete Table 22.4.

Table 22.4 Cultural Eutrophication		
Number of Condominiums	**Phosphorus Added**	**Increase in Algal Population**
10	1 kg	30%
20		
30		
40		
50		

Conclusion: Studying the Effects of Cultural Eutrophication

- Assume that phosphorus is the only nutrient that will cause an increase in the algal population and that *Daphnia* is the only type of zooplankton available to feed on the algae. How many condominiums would you allow the developer to build? _____

  _____

- What other possible impacts could condominium construction have on the condition of the lake?

  _____

1. What type of population would you expect to be the largest in most ecosystems? _____ Explain. _____

   _____

2. What causes acid rain? _____

3. Why is acid deposition harmful to organisms? _____

   _____

4. Name the type of pollution that results when water from rivers and ponds is used for cooling power plants, and explain why it has detrimental effects. _____

   _____

5. Give an example to show that the pollutants studied in this laboratory can have an effect on the human population. _____

   _____

6. When excess nutrients enter an aquatic ecosystem, long-term effects can result. Why?

   _____

   _____

7. Describe how the cultural eutrophication study supports the hypothesis that a balance of population sizes in ecosystems is beneficial. _____

   _____

8. When pollutants enter an ecosystem, they have far-ranging effects. Use acid rain and a food chain to support this statement. _____

   _____

   _____

9. Contrast species composition with species diversity of ecosystems. _____

   _____

10. Suppose that, among sunflower seeds, a particular variety can germinate despite acidic conditions. What do you predict about the survival of that sunflower variety in today's acidic environment compared to the rest of the population? _____

    _____

Metric System

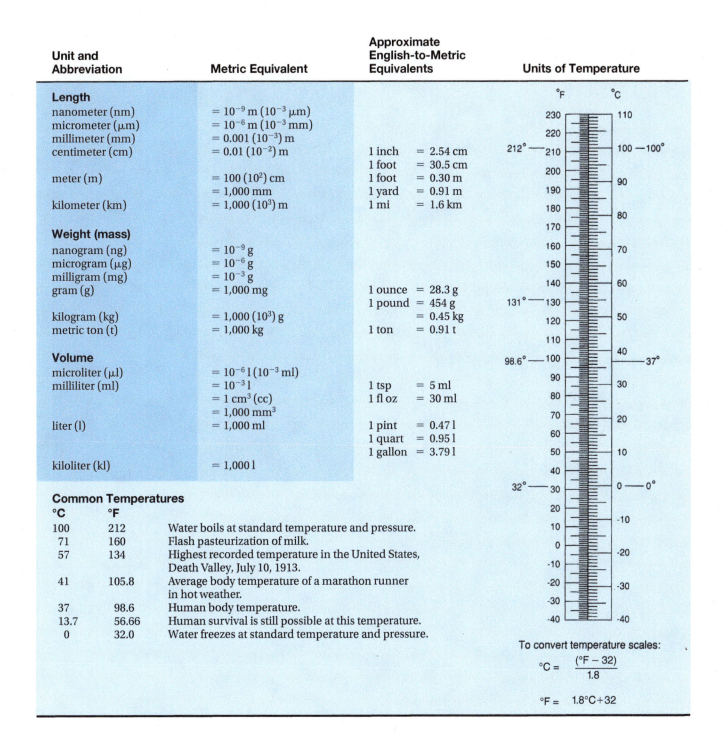

Unit and Abbreviation	Metric Equivalent	Approximate English-to-Metric Equivalents	Units of Temperature

Length

nanometer (nm)	$= 10^{-9}\,\mathrm{m}\ (10^{-3}\,\mu\mathrm{m})$
micrometer (μm)	$= 10^{-6}\,\mathrm{m}\ (10^{-3}\,\mathrm{mm})$
millimeter (mm)	$= 0.001\ (10^{-3})\,\mathrm{m}$
centimeter (cm)	$= 0.01\ (10^{-2})\,\mathrm{m}$
meter (m)	$= 100\ (10^{2})\,\mathrm{cm}$
	$= 1{,}000\,\mathrm{mm}$
kilometer (km)	$= 1{,}000\ (10^{3})\,\mathrm{m}$

English-to-Metric (Length):
1 inch = 2.54 cm
1 foot = 30.5 cm
1 foot = 0.30 m
1 yard = 0.91 m
1 mi = 1.6 km

Weight (mass)

nanogram (ng)	$= 10^{-9}\,\mathrm{g}$
microgram (μg)	$= 10^{-6}\,\mathrm{g}$
milligram (mg)	$= 10^{-3}\,\mathrm{g}$
gram (g)	$= 1{,}000\,\mathrm{mg}$
kilogram (kg)	$= 1{,}000\ (10^{3})\,\mathrm{g}$
metric ton (t)	$= 1{,}000\,\mathrm{kg}$

English-to-Metric (Weight):
1 ounce = 28.3 g
1 pound = 454 g
       = 0.45 kg
1 ton = 0.91 t

Volume

microliter (μl)	$= 10^{-6}\,\mathrm{l}\ (10^{-3}\,\mathrm{ml})$
milliliter (ml)	$= 10^{-3}\,\mathrm{l}$
	$= 1\,\mathrm{cm}^3\ (\mathrm{cc})$
	$= 1{,}000\,\mathrm{mm}^3$
liter (l)	$= 1{,}000\,\mathrm{ml}$
kiloliter (kl)	$= 1{,}000\,\mathrm{l}$

English-to-Metric (Volume):
1 tsp = 5 ml
1 fl oz = 30 ml
1 pint = 0.47 l
1 quart = 0.95 l
1 gallon = 3.79 l

Common Temperatures

°C	°F	
100	212	Water boils at standard temperature and pressure.
71	160	Flash pasteurization of milk.
57	134	Highest recorded temperature in the United States, Death Valley, July 10, 1913.
41	105.8	Average body temperature of a marathon runner in hot weather.
37	98.6	Human body temperature.
13.7	56.66	Human survival is still possible at this temperature.
0	32.0	Water freezes at standard temperature and pressure.

To convert temperature scales:

$$°C = \frac{(°F - 32)}{1.8}$$

$$°F = 1.8°C + 32$$

Credits

Text and Line Art Credits

Lab 1
Laboratory adapted from Kathy Liu, "Eye to Eye with Garden Snails."

Figure 16.12
After Carolina Biological Supply.

Figure 22.2
Drawing by Kristine A. Kohn. From Harriett Stubbs' "Acid Precipitation Awareness Curriculum Materials in the Life Sciences." *The American Biology Teacher* (1983), 45(4), 221. With permission from the National Association of Biology Teachers.

Photo Credits

Laboratory 1
Page 1(pillbug): © James H. Robinson/Science Source; **1.1:** © National Cancer Institute/Science Source.

Laboratory 2
Figure 2.1: © Hill Street Studios/Harmik Nazarian/Getty RF; **2.2:** © Bruce M. Johnson; **2.5:** © B.A.E. Inc./Alamy RF; **2.6(both):** © Richard Hutchings; **2.8:** © moodboard/Alamy RF.

Laboratory 3
Figure 3.2a: © Michael Ross/Science Source; **3.2b:** © CNRI/SPL/Science Source; **3.2c:** © Steve Gschmeissner/Science Source; **3.3-3.4:** © Leica Microsystems, Inc.; **3.6:** © Dr. Gopal Murti/Science Source; **3.7:** © Ted Kinsman/Science Source; **3.8:** © M.I.(Spike) Walker/Alamy.

Laboratory 4
Figure 4.3(both): © Ray F. Evert, University of Wisconsin; **4.7(all):** © David M. Phillips/Science Source; **p. 40(plasmolysis):** © Alfred Owczarzak/Biological Photo Service.

Laboratory 5
Figure 5.2: © McGraw-Hill Education/Jill Braaten, photographer and Anthony Arena, Chemical Consultant.

Laboratory 7
Figure 7.1: © Science Photo Library/Getty Images; **7.3:** © Andrew Syred/Science Source; **7.5(prophase, anaphase, telophase):** © Ed Reschke; **7.5(metaphase):** © Ed Reschke/Photolibrary/Getty Images; **7.6(top):** © Steve Gschmeissner/Science Source; **7.6(bottom):** © Steve Gschmeissner/SPL/Getty RF; **7.7(all):** © Kent Wood/Science Source; **7.8:** © Biophoto Associates/Science Source.

Laboratory 8
Figure 8.1: © Thierry Berrod, Mona Lisa Production/Science Source;
8.3-8.4(all): © Ed Reschke; **p. 78(dogs):** © American Images, Inc./Getty Images.

Laboratory 9
Figure 9.4: © Evelyn Jo Johnson; **9.5:** © Carolina Biological Supply Company/Phototake.

Laboratory 11
Figure 11.5(all): © CNRI/SPL/Science Source.

Laboratory 13
Page 144(Gram-stained bacteria): © ASM/Science Source; **13.1:** © Dr. Kari Lounatmaa/Science Source; **13.2a:** © Alfred Pasieka/Science Source; **13.2b:** © Science Photo Library/Getty RF; **13.2c:** © SciMAT/Science Source; **13.3a:** © Steven R. Spilatro, Marietta College, Marietta, OH; **13.3b:** © Biophoto Associates/Science Source; **13.4:** © M.I. Walker/Science Source; **13.5:** © R. Knauft/Science Source; **13.7b:** © M.I. Walker/Science Source; **13.8(both):** © Manfred Kage/Science Source; **13.10:** © Jan Hinsch/Science Source; **13.12c:** © Ed Reschke; **13.13b:** © M.I. (Spike) Walker/Alamy; **13.15a:** © Biophoto Associates/Science Source; **13.15b:** © Kenneth H. Thomas/Science Source; **13.15c:** © imagebroker/Alamy RF; **13.15d:** © Matt Meadows/Photolibrary/Getty Images; **13.16a:** © Gary T. Cole/Biological Photo Service; **13.17(bread):** © Jules Frazier/Getty RF; **13.17(zygospore):** © Ed Reschke/Photolibrary/Getty Images; **13.18(both):** © Carolina Biological Supply Company/Phototake; **13.20a:** © Biophoto Associates/Science Source; **13.20b:** © John Hadfield/SPL/Science Source; **13.20c:** © P. Marazzi/SPL/Science Source.

Laboratory 14
Figure 14.1: © Stephen P. Lynch; **14.3(left):** © Bob Gibbons/Science Source; **14.3(right):** © Kingsley Stern; **14.6:** © Steve Solum/Photoshot; **14.8:** © McGraw-Hill Education/Carlyn Iverson, photographer; **14.9:** © Carolina Biological Supply/Phototake; **14.11a:** © Creatas/Jupiterimages; **14.11b:** © Kathy Merrifield/Science Source; **14.11c:** © Evelyn Jo Johnson; **14.13a:** © J.R. Waaland/Biological Photo Service; **14.13b:** © Ed Reschke; **14.14:** © Carolina Biological Supply/Phototake.

Laboratory 15
Figure 15.4a: © Carolina Biological Supply/Phototake; **15.5:** © N.C. Brown Center for Ultrastructure Studies, SUNY, College of Environmental Science & Forestry/Dr. Wilfred A. Cote; **15.6a:** © Ray F. Evert, University of Wisconsin; **15.6b:** © Carolina Biological Supply Company/Phototake;
15.7: © John Selier Virginia Tech Forestry Department; **15.8:** © Carolina Biological Supply/Phototake; **15.9b:** © J.R. Waaland/Biological Photo Service.

Laboratory 16
Figure 16.1: © Renaud Visage/Photographer's Choice RF/Getty RF; **16.3(land snail):** © Georgette Douwma/Science Source; **16.3(nudibranch):** © Kenneth W. Fink/Bruce Coleman/Photoshot; **16.3(octopus):** © Alex Kerstitch/Bruce Coleman/Photoshot; **16.3(nautilus):** © Douglas Faulkner/Science Source; **16.3(scallop):** © ANT Photo Library/Science Source; **16.3(mussels):** © Michael Lustbader/Science Source; **16.5b-16.6b:** © Ken Taylor Wildlife Images; **16.7a:** © Kim Taylor/Bruce Coleman/Photoshot; **16.7b:** © Borut Furlan/WaterFrame/Getty Images; **16.7c:** © NHPA/Photoshot; **16.7d:** © L. Newman & A. Flowers/Science Source; **16.9:** © Ken Taylor Wildlife Images; **16.10(dragonfly):** © Thomas Shahan/Flickr Select/Getty Images; **16.10(louse):** © Alastair MacEwen/Oxford Scientific/Getty Images; **16.10(wasp):** © James H. Robinson/Science Source; **16.10(butterfly):** © Wind Home Photography/Flickr/Getty RF; **16.10(beetle):** © George Grall/National Geographic/Getty Images; **16.10(leafhopper):** © Whitehead Images/Alamy RF; **16.16a:** © Ihoko Saito/Toshiyuki Tajima/Getty RF; **16.16b:** © image100/PunchStock RF; **16.16c:** © Rod Planck/Science Source; **16.16d:** © Suzanne and Joseph Collins/Science Source; **16.16e:** © Bill Horn/Acclaim Images RF; **16.16f:** © Craig Lorenz/Science Source.

Laboratory 17
Figure 17.3b,17.6: © Ken Taylor/Wildlife Images.

Laboratory 18
Figure 18.2: © Ed Reschke.

Laboratory 20
Figure 20.6: © MedImage/Science Source; **20.8, 20.11:** © McGraw-Hill Education/Carlyn Iverson, photographer.

Laboratory 21
Figure 21.2(all): © Dr. J. Timothy Cannon; **21.4:** © Kage Mikrofotografie/Phototake; **21.6:** © P.H. Gerbier/SPL/Science Source; **21.7a(ommatidia):** © PIR/Science Source; **21.7a(housefly):** © L. West/Bruce Coleman/Photoshot; **21.7b:** © Douglas Faulkner/Science Source.

Laboratory 22
Figure 22.2: © William Amos/Photoshot.

Index

Note: Page numbers followed by f refer to figures; page numbers followed by t refer to tables.

Brain stem, 270
Bronchi, 218
Brown algae, 150, 151f
Bryophytes, 167
Buffer, 42
Bulbourethral glands, 260
Butterflies, metamorphosis in, 206

C

Cambrian period, 126, 127t
Cancer, cell cycle and, 64
Capsule, 144, 145f
Carapace, in crayfish, 202, 202f
Carbon cycle, 61, 61f
Carbon dioxide, in photosynthesis, 53, 53f
Carbon dioxide uptake, 60
Carboniferous period, 126, 127t
Cardiovascular system
    arteries and veins in, 255, 255f
    description of, 252
    observation of, 252–254, 252f, 253f
    pulmonary circuit in, 252, 254
    systemic circuit in, 252, 254
    in vertebrates, 256, 256f
Carotenes, 54
Carriers, 115
Cartilaginous fishes, 208
Catalase
    function of, 45–47
    pH and, 50
Cell cycle
    cancer and, 64
    description of, 64, 64f
    M stage and, 65
    S stage and, 65, 65f
Cell plate, 71, 71f
Cells. *See also* Animal cells; Plant cells
    daughter, 65–67, 71, 71f, 74, 75f, 77f
    microscopic observation of, 19, 19f, 28, 28f
    overview of, 31
    parent, 65, 66
    tonicity of, 39–40, 39f
Cell structure/function
    animal and plant, 32, 32t, 33f, 34f
    diffusion and, 36–37, 36f, 37t
    observation of, 35, 35f
    osmosis and, 38–41, 38f–40f
    overview of, 31
    pH and, 42–43, 42f
Cell theory, 31

Cellular reproduction
    animal cell mitosis and cytokinesis and, 66–68, 66f–68f
    cell cycle and, 64–65, 64f, 65f
    overview of, 63, 63f
    plant cell mitosis and cytokinesis and, 68–71, 69f, 71f
Cellular respiration
    carbon cycle and, 61
    photosynthesis and, 57
Cell wall, 32t, 144, 145f
Celsius scale, 16, 16f
Cenozoic era, 126, 127t
Centimeter, 10, 10t
Central nervous system (CNS)
    description of, 267, 267f
    human brain and, 270
    sheep brain and, 268, 269f, 270
    spinal cord and, 272, 272f
    vertebrate brain comparisons and, 271, 271f
Central tendon, 216
Centrioles, 32t, 65t
Centromeres, 65, 65t, 66, 66f
Centrosome, 32t
Cephalothorax, in crayfish, 202, 202f
Cerebellum, 268, 271
Cerebral hemispheres, 268
Chara, 165, 165f
Chemoreceptors, 284
Chimpanzees
    chromosomal and genetic data of, 138
    skeleton of, 133–134, 134f
    skull features of, 135, 135f
Chlorella, 292, 293
Chlorophyll *a*, 54
Chlorophyll *b*, 54
Chloroplast
    description of, 32t
    observation of, 35, 35f
Choanoflagellate, 194
Chordae tendineae, 253, 253f
Chordates, 194, 207, 207f
Chromatids, 65, 65t, 66, 66f, 73, 74
Chromatin, 65
Chromatography, 54–55, 54f
Chromosome inheritance, method to determine, 119–120, 120f
Chromosomes
    autosomal, 112
    function of, 65t
    homologue, 74, 74f, 76
    human, 79, 112
    meisosis and, 73
    sex, 93, 120, 120f

Cilia, 32t, 154
Circulatory system, of clams, 197
Clams
    anatomy of, 196–197, 196f, 198f
    bivalve, 198f
    description of, 196
Claspers, in grasshoppers, 204
Club fungi, 159–160, 159f
Club mosses, 169, 169f
Coccus, 147, 147f
Cochlea, 281f, 282
Codon, 104
Cold, sensory receptors for, 283
Colon, in fetal pig, 222
Colonies
    daughter, 150, 151f
    description of, 146, 150
Color blindness, 117, 117f
Colpoda, 288f
Common ancestors, 125
Common descent, 125
Comparative anatomy
    evolution and, 131–137, 131f, 133f–136f
    homologous structures and, 125
Compound, 20
Compound eyes, 202, 204, 274, 275f
    in crayfish, 202
    in grasshoppers, 204
Compound light microscopes
    function of, 20, 24
    method to focus, 26
    parts of, 24–25, 24f
    transmission electron microscope vs., 21f
Concentration, enzyme activity and, 49
Conclusion, 3, 3f, 6–7
Cone cells, 276, 277t
Cones, 173–175, 173f–175f
Conifers, 173, 173f
Conjugation, 150, 150f
Conjugation pilus, 144, 145f
Contractile vacuoles, 154
Control culture, 288
Cork, 189, 189f
Corn kernels, 86–87, 86f
Corpus callosum, 268
Cortex, in plants, 185, 185f, 187, 187f, 189, 189f
Cowper's glands, 260
Cranial nerves, 272
Crayfish, anatomy of, 202–203, 202f, 203f
Crenation, 39, 39f
Cro-Magnons, 136

Eyes
accommodation of, 276, 279, 279f
blind spot of, 278, 278f
camera-type, 274, 275f
compound, 202, 204, 274, 275f
of crayfish, 202, 202f
of fly, 275f
function of, 274
of grasshopper, 204
human, 276, 277f, 277t
invertebrate, 274, 275f

F

Fahrenheit scale, 16
Fat digestion, 229–230, 229f, 232
Female reproductive system, 263, 263t, 264f, 265f
Ferns, 167, 170–171, 170f, 171f
Fertilization, 177, 177f
Fetal pig
abdominal and thoracic cavity of, 216, 217f, 218, 220, 221f, 222
arteries and veins of, 211, 255f
cardiovascular system of, 251–256, 251f–253f, 255f, 256f
digestive system of, 220, 221f, 222, 257
dissection of, 211, 211f
external anatomy of, 213f
female reproductive system in, 263, 263t, 264f, 265f
internal anatomy of, 219f
male reproductive system in, 260, 260t, 261f, 262, 262f
neck region of, 218
oral cavity and pharynx of, 214–215, 214f, 215f
respiratory system of, 257
storage of, 222, 265
urinary system of, 258f, 259
Fishes
brain in, 271
cardiovascular system of, 256f
evolution of, 208
lateral line system of, 280f
Flagella, 32t, 144, 145f, 153, 154, 154f
Flowering plants. See also
Angiosperms; Plants
anatomy of, 181
description of, 163, 163f
life cycle of, 172, 176
monocot and eudicot, 183
Flowers, 177–178, 177f
Fly, eye of, 275f

FOIL method, 89
Food diary, 235, 236
Foods, nutrient and energy content of, 237f
Food vacuoles, 154
Foot, of molluscs, 195
Forebrain, 271
Forelimbs, of vertebrates, 131, 131f, 132
Forewings, of grasshoppers, 204
Fossil record, 126, 127t
Fossils
description of, 125, 126
invertebrate, 129
plant, 130, 130f
vertebrate, 129
Frond, 170
Frontal lobe, 268
Fucus, 151, 152f
Fume hood, 55
Fungi
black bread mold, 158, 158f, 159f
club, 159–160, 159f
description of, 157
human disease and, 160

G

Gallbladder
of fetal pig, 219f, 220, 222
function of, 226f
Gametes, 63, 73
Gametogenesis, 63, 73, 74
Gametophyte, 166, 166f, 167
Gammarus, 291, 291f
Gel electrophoresis, 106–107, 107f, 109, 109f
Genes
forms of, 111
mutated, 97
Genetic counseling
description of, 119
to determine chromosome inheritance, 119–120, 120f
to determine pedigree, 121–123, 121f, 122f
Genetic disorders. *See also*
Inheritance
autosomal, 115–116, 115f
color blindness as, 117, 117f
detection of, 107–109, 108f, 109f
inheritance of, 114–118, 115f, 117f
X-linked, 116–117, 117f
Genetics
multiple alleles and, 118–119

patterns related to, 114, 114f
principles of, 111
Genitals, in mammals, 212
Genomic sequencing
function of, 108
for sickle-cell disease, 108–109, 108f, 109f
Genotype
description of, 83, 83f, 84, 111
method to determine, 112–113
phenotype vs., 111, 111f
Geological timescale
dating with, 128
divisions of, 126
limitations of, 128, 128f
table of, 127t
Gill mushroom, 159, 159f
Gills, of clams, 197
Gloeocapsa, 148, 148f
Glottis, of fetal pig, 215, 215f, 257
Glucose, 53
Golden-brown algae, 153, 153f
Golgi apparatus, 32t
Gram, 13
Grasshoppers
anatomy of, 203–204, 203f, 205f
ears of, 280
metamorphosis in, 206
Gray matter, 272, 272f
Green algae, 150, 150f, 151, 151f, 165, 165f, 292–293
Green glands, in crayfish, 202
Green light, in photosynthesis, 58–59
Gross photosynthesis, 57
Gymnosperms
description of, 172, 173, 173f
life cycle of, 173–175, 173f–175f

H

Hair cells, 282
Hairline, 121
Haploid (n), 74
Hard palate, 214, 214f
Hay infusion cultures, 288–289, 288f
Heart
external view of, 252–253, 252f
of fetal pig, 218
function of, 251
internal view of, 253–254, 253f
Heat, sensory receptors for, 283
Hemolysis, 39, 39f
Hemophilia, 117
Herbaceous plants. *See also* Plants
description of, 181

stems in, 187, 187f
Heterozygous alleles, 111
Hindbrain, 271
Hind wings, of grasshoppers, 204
Hinge ligament, 197
Hominids, skulls of, 136
Homologous structures, 125, 131, 131f
Homologues, 74, 76, 76f
Homo sapiens, 126, 136. *See also* Humans
Homozygous dominant, 83
Homozygous recessive, 83
Homozygous recessive alleles, 111
Human Genome Project, 107
Humans. *See also* Cellular reproduction
   anatomy of, 223f
   autosomal traits in, 113, 113t
   body composition for, 246, 247t, 248t, 249
   brain in, 270
   chemoreceptors in, 284
   chromosomes of, 79, 112
   energy requirements for, 235–249 (*See also* Energy requirements)
   evolution of, 136, 136f
   eyes in, 276, 277f, 277t
   ideal weight for, 244–249, 244t, 245t, 247t, 248t
   life cycle of, 79–81, 79f, 81f
   percentage fat estimates for, 246, 247t, 248t
   reproductive system in, 262, 262f (*See also* Reproductive system)
   skeleton of, 133–134, 134f
   skin in, 282–283, 282f
   skull features of, 135, 135f
Huntington disease, 115
Hydrogen peroxide, 46, 46f
Hypertonic solution, 39, 39f
Hyphae, 157, 157f
Hypothalamus, 268
Hypothesis
   description of, 2, 3f
   formulation of, 2, 6
Hypotonic solution
   description of, 39, 39f
   plant cells in, 40

I

Immune system, 139
Incus, 282
Indusium, 171

Inguinal canal, 260
Inheritance. *See also* Genetic disorders
   autosomal, 115–116, 115f
   chromosome, 119–120, 120f
   of color blindness, 117, 117f
   of genetic disorders, 114–118, 115f, 117f
   one-trait crosses and, 84–87, 84f–87f
   two-trait crosses and, 88–92, 89f–91f, 92t
   x-linked crosses and, 93–94, 93f, 94t
Inner ear, 281t
Insects
   description of, 193, 203
   diversity of, 203f
   grasshoppers as, 203–204, 203f, 205f
   metamorphosis in, 206, 206f
Interkinesis, 74f, 75
Interneuron, 273
Internode, in plants, 182, 182f
Interphase, in cell division, 63, 64
Intestines
   of clams, 197
   of fetal pig, 220, 221f
Inversion, of microscopic image, 26
Invertebrate chordates, 207
Invertebrate fossils, 129
Invertebrates
   arthropods as, 201–203
   clams as, 196–197, 196f, 198f
   description of, 129, 193, 195
   eyes in, 274, 275f
   grasshoppers as, 203–204, 203f, 205f
   insect metamorphosis in, 206, 206f
   molluscs as, 195–196, 195f
   squids as, 199, 200f
Isotonic solution, 39, 39f

J

Jacob syndrome, 120
Jawless fishes, 208

K

Karyotypes, 119
Kelps, 151
Kidneys, 259, 259f
Klinefelter syndrome, 120

Knee-jerk reflex, 274, 274f
K-T extinction, 128

L

Labial palps, 197
Laminaria, 151, 152f
Lancelets, 207
Large intestine
   of fetal pig, 220, 221f, 222, 257
   function of, 226f
Larynx, of fetal pig, 218, 257
Lateral bud, 182, 182f
Latex gloves, 212, 255
Law of independent assortment, 88
Law of segregation, 84
Leaf scar, 188, 188f
Leaf vein, 190, 191, 191f
Lean body weight (LBW), 246, 247t, 248t, 249
Leaves
   anatomy of, 191, 191f
   function of, 181, 190
   stomata in, 190, 190f
Length, measurement of, 10–11, 10t, 11f, 17f, 17t
Life cycle
   of flowering plants, 176–178, 176f–178f
   of humans, 79–81, 79f, 81f
   of plants, 166, 166f, 167f, 170, 170f
   of seed plants, 173–175, 173f–175f
Light, in photosynthesis, 56–59, 56f–58f
Light micrographs, 20
Liter, 14
Liver
   of fetal pig, 220, 221f
   function of, 226f, 229
Liverworts, 167
Lobe-finned fishes, 208
Longitudinal section, 19, 19f
Lungs, of fetal pig, 218
Lycophytes, 169, 169f
Lympocytes, 20, 20f
Lysosome, 32t

M

Macrocystis, 152f
Magnification, total, 27
Malaria, 154
Male reproductive system, 260, 260t, 261f, 262, 262f

Malleus, 282
Mammals. *See also* Humans; *specific animals*
    abdominal and thoracic cavity of, 216, 217f, 218, 220, 221f, 222
    anatomy of human, 223, 223f
    brain in, 271
    cardiovascular system of, 251–256, 251f–253f, 255f, 256f
    digestive system of, 220, 221f, 222, 257
    exterior anatomy of, 212–213, 213f
    female reproductive system in, 263, 263t, 264f, 265f
    internal anatomy of, 218, 219f
    male reproductive system in, 260, 260t, 261f, 262, 262f
    neck region of, 218
    oral cavity and pharynx of, 214–215, 214f, 215f
    respiratory system of, 257
    urinary system in, 259, 259f
Mammary glands, 212
Mantle
    of clams, 197
    of molluscs, 195
Mass extinction, 127t, 128, 128f
Measurement, metric system, 9–17
Medulla oblongata, 270
Megasporangia, 172
Meiosis
    crossing-over in, 77
    description of, 63, 73, 74
    in human life cycle, 79
    mitosis vs., 80, 81f
    phases of, 74, 74f, 75f
    production of variation during, 76–78, 76f–78f
    in seed plants, 172
Meiosis I
    description of, 73
    phases of, 74, 74f
Meiosis II
    description of, 73, 74
    phases of, 74, 75, 75f
Mendel, Gregor, 83, 84, 88
Meniscus, 14, 15f
Meristem, 182
Mesenteries, 220
Mesophyll, 191
Mesothorax, of grasshopper, 204
Mesozoic era, 127t
Messenger RNA (mRNA), 97, 102
Metabolism, 45
Metamorphosis, 206, 206f
Metaphase, 74, 74f, 75f, 77

Metathorax, of grasshopper, 204
Meter, 10, 10t
Meterstick, 10, 10t, 11t
Methylene blue, 28
Metric system
    description of, 9, 9f
    length in, 10–11, 10t, 11f, 17f
    temperature, 16, 17f
    volume in, 14–15, 15f, 17f
    weight in, 13–14, 13f, 14f, 17f
Micrographs, comparative, 20f, 21
Micrometer, 10, 10t, 11, 11f
Microscopes
    compound light, 24–25, 24f
    electron, 21, 21f
    inversion of image by, 26
    light, 20, 20f
    rules for use of, 21
    stereomicroscope, 19, 22–23, 22f
    total magnification in, 27, 27t
    types of, 19
Microscopic observation
    of *Euglena*, 29, 29f
    of human epithelial cells, 28, 28f, 31
    of onion epidermal cells, 28, 28f
    wet mount preparation for, 27, 27f
Microsporangia, 172
Midbrain, 270, 271
Middle ear, 281t
Milliliter, 14
Millimeter, 10, 10t, 11
Millions of years (MYA), 128
Millogram, 13
Mitochondrion, 32t
Mitosis
    animal cell, 66–68, 66f–68f
    cell cycle and, 65
    description of, 63, 66
    in human life cycle, 79
    meiosis vs., 80, 81f
    plant cell, 68–71, 69f, 71f
    structures associated with, 65t
Mitotic spindle, 66, 66f
Mitral valve, 253, 253f
Molecular data, evolution and, 138–139, 138f–140f
Molecular genetics, 97
Molluscs
    classes of, 195, 195f
    description of, 193, 195
Monocot plants
    description of, 183, 183f
    stems in, 187, 187f
Monohybrid cross, 84, 85f
Mosses, 167–169, 167f, 169f

Moths, metamorphosis in, 206
Motor neuron, 273
Mouth, 226f
Multiple alleles, 118
Mushrooms, 159–160, 159f
Mycelium, 157

N

Nanometer, 10, 10t, 11, 11f
Neck region, of fetal pig, 218
Negative control, 2, 226
Nephrons, 259, 259f
Nereocystis, 152f
Nervous system
    central, 267, 267f, 268, 269f, 270–272, 271f, 272f
    overview of, 267, 267f
    peripheral, 272–274, 272f–274f
Net photosynthesis, 56
Neurofibromas, 115
Neurofibromatosis (NF), 115
Nipples, in mammals, 212
Node, in plants, 182, 182f, 188, 188f
Notochord, 207
Nuclear envelope, 32
Nucleoid, 144, 145f
Nucleolus, 32t, 65t
Nucleoplasm, 32
Nucleotides, 98, 101, 101f
Nucleus, cell, 32, 32f, 65t

O

Observation, 2, 3f, 4
Occipital lobe, 268
Ommatodoa, 274
One-trait crosses, 84–87, 84f–87f
Oogenesis, 79
Oral cavity, in mammals, 214
Oral thrush, 160, 160f
Ordovician period, 126, 127f
Organelles, 63
Oscillatoria, 148, 148f
Osmosis
    description of, 38
    observation of, 38, 38f
    tonicity in cells and, 39–40, 39f
Outer ear, 281t
Ovaries, 263, 263t, 264f
Oviducts, 263, 263t
Ovipositors, 204
Ovule, 172
Oxygen-deprived culture, 289

P

Palate, 214, 214f
Paleontologists, 126, 136
Paleozoic era, 126, 127t
Palisade mesophyll, 191
Pancreas
     of fetal pig, 219f, 220, 222
     function of, 226f
Pancreatic amylase, 231–232
Pancreatic lipase, fat digestion by,
     229–230, 229f
Paper chromatography, 54–55, 54f
Paramecia, 154, 154f, 288f
Parent cells, 65, 66
Parfocal, 26
Parietal lobe, 268
Patellar tendon, 274, 274f
Paternity, blood type to determine,
     118
Pathogenic bacteria, 144–145, 144f,
     145f
Pedigree, 121–123, 121f, 122f
Penis, 213, 260
Pepsin, protein digestion by, 227–228,
     227f
Peptidoglycan, 144, 145f
Pericardial cavity, of fetal pig, 218
Pericardium, of clams, 197
Pericycle, in roots, 185, 185f
Periods, 126, 127t
Peripheral nervous system (PNS),
     267, 267f, 272–274, 272f–274f
Peritoneal cavity, 259
Peritoneum, 220, 259
Peroxisome, 32t
pH
     cells and, 42–43
     description of, 42
     enzymatic activity and, 50
Pharyngeal pouches, 132, 133f
Pharynx
     function of, 226f
     in mammals, 214, 215, 215f
Phenotype
     description of, 83, 83f, 111
     genotype vs, 111, 111f
     inheritance of, 114, 115f
     X-linked recessive, 121
Phenotypic ratio, 84
Phenylketonuria (PKU), 116
Philodina, 288f
Phloem, 185, 185f, 188, 189, 189f
Photomicrographs
     description of, 20, 20f
     limitations of, 31

Photosynthesis
     action spectrum for, 58, 58f
     carbon cycle and, 61, 61f
     carbon dioxide uptake and, 60
     description of, 53, 53f, 56
     equation for, 53
     gross, 57
     net, 56
     plant pigments and, 54–55, 54f
     rate of, 57
     solar energy and, 56–59, 56f–58f
pH scale, 42, 42f
Phylum chordata, 195
Physical activity, average daily energy
     required for, 239, 240t
Pig. *See* Fetal pig
Pili, 144, 145f
Pillbugs, 1–2, 1f, 4
Pine cones, 174, 174f, 175, 175f
Pipet, 15, 15f
Pith, in plants, 187, 187f, 189, 189f
Placenta, in mammals, 212
Plant cell mitosis
     description of, 68
     observation of, 69, 69f
     phases of, 69–70
Plant cells
     cytokinesis in, 68–71, 69f, 71f
     structure of, 32, 32t, 34f, 35, 35f
     tonicity in, 40, 40f, 41
Plant evolution
     seedless plants and, 167–171, 167f,
          169f–171f
     seed plants and, 172–179,
          172f–177f
Plant fossils, 130
Plant pigments, photosynthesis and,
     54–55, 54f
Plants
     alteration of generations of, 166,
          166f
     evolution of, 163, 164f, 165, 165f
     flowering, 163, 163f
     groups of, 130f, 181
     herbaceous, 181
     leaves of, 190, 190f, 191f
     life cycle of, 166, 166f, 167f, 170,
          170f
     overview of, 181
     roots of, 184–185, 184f, 185f
     root system and shoot system in,
          182–183, 182f, 183f
     seed, 172–179, 172f–179f (*See also*
          Seed plants)
     seedless, 167–171, 167f, 169f–171f
          (*See also* Seedless plants)

stems of, 187–189, 187f–189f
     system transport in, 186, 186f
     woody, 181
Plasma membrane
     diffusion across, 36–37, 37f, 37t
     function of, 31, 32f, 144, 145f
Plasmodium vivax, 154
Pleural cavity, of fetal pig, 218
Poles, of spindle, 66, 66f
Pollen cones, 174–175, 174f–175f
Pollen grain, 172
Pollen tube, 177
Pollination, 172, 177
Pollution
     cultural eutrophication and,
          292–294
     description of, 287
     effect on *Gammarus*, 291, 291f
     effect on hay infusion culture,
          288–289, 288f
     effect on seed germination, 290
     types of, 287
Polypeptides, 105
poly-X syndrome, 120
Pond water, protists found in, 155,
     156f
Pons, 270
Positive control, 226
Posture, human vs. chimpanzee,
     133–134, 134f
Precambrian time, 126, 127t
Precipitate, 139
Primary growth, in plants, 181
Primates, molecular data
     among, 138
Products, in chemical reaction, 45
Prokaryotes, groups of, 143
Prophase, 74f, 75f, 77
Prostate gland, 260
Proteins, digestive system and,
     227–228, 227f
Protein similarity, evolution and, 139,
     139f, 140f
Protein synthesis
     description of, 102
     DNA and, 102–105, 103f–105f
     transcription in, 102, 103, 103f
     translation in, 102, 104–105, 104f,
          105f
Prothallus, 171
Prothorax, of grasshopper, 204
Protists
     algae as, 149–153, 150f–153f
     description of, 149
     protozoans as, 154–156,
          154f–156f

Protoslo, 155
Protostomes, 194
Protozoans
    description of, 154
    observation of, 155–156, 155f, 156f
Pseudopods, 154
P-T extinction, 128
Pulmonary circuit, 254
Pulmonary semilunar valve, 253, 253f
Pulmonary trunk, 252f, 253
Pulmonary veins, 252f, 253
Punnett squares, 84, 84f, 114, 114f
Pyloric sphincter, 257
Pyrenoids, 150

R

Radial symmetry, 194
Radula, of squid, 199
Ray-finned fishes, 208
Rays, in plants, 189
Reactants, 45
Receptacles, 151
Recessive alleles, 111, 116
Rectum
    of fetal pig, 222, 257
    function of, 226f
Red algae, 151, 152
Red blood cell, tonicity and, 39, 39f
Reflexes, 273–274, 273f
Refraction, 276
Renal cortex, 259, 259f
Renal medulla, 259, 259f
Renal pelvis, 259, 259f
Reproduction. See Cellular
    reproduction; Sexual
    reproduction
Reproductive system. See also Sexual
    reproduction
    female, 263, 263t, 264f, 265f
    male, 260, 260t, 261f, 262, 262f
Reptiles
    brain in, 271
    cardiovascular system of, 256f
    description of, 208
Resolution, 21
Respiration, in insects, 203
Respiratory system, of fetal pig, 257
R_f (ratio-factor) values, 55
Rhizome, 170
Ribonucleic acid. See RNA
    (ribonucleic acid)
Ribosomal RNA (rRNA), 97, 102
Ribosome, 32t, 105f
Ribosomes, 144, 145f

Ringworm, 160, 160f
RNA (ribonucleic acid)
    structure of, 97, 101–102, 101f,
        102t
    transcription and, 102, 103
    translation and, 104–105, 104f,
        105f
Rockweed, 151, 152f
Rod cells, 276, 277t
Roly-poly bugs. See Pillbugs
Romalea, 204, 205f
Root apical meristem, 188
Root hairs, 184
Roots
    function of, 181, 182, 182f
    organization of, 184–185, 184f,
        185f
Rough endoplasmic reticulum, 32t

S

Salivary glands, 226f
Saprotrophs, 157
Scanning electron micrographs, 20f
Scanning electron microscopes,
        19, 21. See also
        Stereomicroscopes
Scientific method, 2–3, 3f
Scientific theory, 3, 3f
Scrotal sacs, 260
Seaweed, 151–152, 152f
Secondary growth, in plants, 181
Secondary xylem, 188
Seed germination, 290
Seedless plants. See also Plants
    description of, 167
    ferns as, 167, 170–171, 170f, 171f
    mosses as, 167–169, 167f, 169f
Seed plants. See also Plants
    angiosperms as, 172, 176–179,
        176f–179f
    description of, 172
    dispersal in, 172, 172f
    gymnosperms as, 172–175,
        173f–175f
    life cycle of, 176–179, 176f–179f
Semen, 260
Seminal receptacles, 202
Seminal vesicles, 260
Sense strand, 103
Sensory neurons, 273
Sensory receptors
    in skin, 282–283, 282f
    for temperature changes, 283
Septa, 157

Sex chromosomes
    anomalies of, 120, 120f
    description of, 93, 112
Sexual reproduction. See also
        Reproductive system
    in black bread mold, 158, 158f,
        159f
    in club fungi, 159f
    by conjugation, 150, 150f
    description of, 73, 73f
    in females, 263, 263t, 264f, 265f
    human life cycle and, 79–81,
        79f, 81f
    in males, 260, 260t, 261f, 262,
        262f
    meiosis in, 74–75, 74f, 75f
    production of variation during
        meiosis in, 76–78, 76f–78f
    of volvox, 150, 151f
Sheep, brain in, 268, 269f, 270
Shoot apical meristem, 188
Shoot system, 182, 182f
Sickle-cell disease
    description of, 108
    detection of, 109, 109f
    genetic sequence for, 108–109,
        108f, 109f
Simple eyes, in grasshoppers, 204
Siphons, in clams, 197
Sister chromatids, 65, 75
Skeletons, chimpanzee vs. human,
        133–134, 134f
Skin, sensory receptors in human,
        282–283, 282f
Skulls
    chimpanzee vs. human, 135, 135f
    hominid, 136
Small intestine
    fat digestion in, 232, 232f
    of fetal pig, 219f, 220, 221f, 222,
        257
    function of, 226f
Smell receptors, 284
Smooth endoplasmic reticulum, 32t
Soft palate, 214, 214f
Solar energy, photosynthesis and, 53,
        53f, 56–59, 56f–58f
Solution, tonicity of, 39, 39f
Sori, 170, 170f, 171f
Species, mass extinction of, 127t,
        128
Species composition, 288
Species diversity, 288
Specific dynamic action (SDA), 235,
        243
Spectrophotometer, 293

X

Xanthophylls, 54
X-linked crosses, 93–94, 93f, 94t
X-linked disorders
    color blindness as, 117, 117f
    description of, 116, 117f
    hemophilia as, 117

X-linked recessive phenotype, 121,
    122
Xylem
    annual ring and, 189, 189f
    description of, 181, 185, 185f
    transport function of, 186, 186f

Z

Zygospore, 158
Zygotes
    description of, 73, 73f, 79
    in ferns, 167, 170
    in seed plants, 172